Plane Talk

Plane Talk

Jack L. Dowd Ph.D.

nutcracker1@hawaiiantel.net

'THIS A WORK OF NON FICTION'

1. AIRFRAME & POWERPLANT MECHANICS AC-12A page18, cylinder
2. ACCEPYABLE METHODS, Techniques and Practitioners EA-43.13—1A & A2 page 271 col. 1
3. AIRFRAME & POWERPLANT MECHANIC Airframe handbook Page 25 col. 1
4. Honolulu Star Bullet November 21, Page 1, col.

This book was printed in the United States of America.

To order additional copies of this book, contact:
Xlibris Corporation
1-888-795-4274
www.Xlibris.com
Orders@Xlibris.com
68792

CONTENTS

SYNOPSIS..11
FORWARD 13
THROUGHTS FROM THE PAST 15
VOCABULARY..17
MECHANIC'S CREED..19
PLANE TALK..21
JET ENGINE OVERHAUL AND MAINTENANCE 23

1   Ma Worldly..31
2   Chief Pilot got Fired, President got hired. 34
3   The Company hires a Supervisor..................................37
4   Tank dry Over speed Engine change. 39
5   Gear will not come down Electrically that's what
     the clutch is for Emergency!! 41
6   Aircraft Engines quit over Ft. Kam, 4 people die.......43
7   Aircraft came into high, replaced right landing gear & right flap. 45
8   Aircraft has no oil pressure right engine on take off from Kauai,
     Engine change..46
9   We got past Kona we declared holiday..........................48
10   Using 207 gallon of fuel, instead of the normal
     250 gallon of fuel to SAVE money. 49
11   Engine runs rough due to Manual Leaning,
     save money spend it on repairs. 52
12   Found crack in the Wing Fitting during X-RAY INSPECTION.....53
13   Mechanic sat on top of engine to look for magneto drop,
     could have been killed. 54
13a 55 has a problem..55
13b The Fog Factor...56

13c To cold for Mainlanders in Hawaii..58
13d Can't fix the airplanes & 9th seat..59
14   Collector Ring Vs short stack exhaust system. 62
15   Manual Leaning is it necessary. 64
16   Chief Pilot gets a PAY RAISE, I want mine.............................67
17   Battery Fire Reported Fire, Replaced Battery. 69
18   Gear problem, down and locks he thinks,
     indication light would not come on 70
19   Pilot left the battery on for 3 hours on Kauai lost power,
     unlocked gear, gear collapsed on landing...............................71
20   Aircraft hit a radio guide wire at KOKO HEAD crater pilot
     and CO-pilot were killed. 72
21   Mechanic filled the engine with water, two bottom cylinder
     removed and water was expelled, ran engine for 10 minutes.........73
22   Nose Wheel collapsed on landing
     at Dillingham Field scrap aircraft, for spar parts. 74
23   Aircraft landed on its nose wheel,
     fuselage look like an accordion, took a week to repair...................75
24   Shimmy Damper fails in Kona..76
25   Mixture control rod came off, not safety..................................77
26   A. C. E. cargo aircraft (DC-4) lost an Engine on take off............78
27   Propeller would not go through a complete cycle,
     link rod pin moved out of position. 79
28   24 aircraft flew this day, 18 of ours, 6 of theirs. 80
29   Aircraft that would not fly...81
30   First wheels up landing the Director Maintenance surprise. 83
31   Second Wheels up landing, he got FIRED. 85
32   Buzzing sound coming from the engine, BEES?.......................87
33   Aircraft ended up in the culvert on Maui,
     change prop, change cylinder 88
34   Number one piston did not move up or down,
     Replace engine or replace link rod Chief Pilot knows. 90
35   One aircraft chew-up the other one, change elevator
     on the plane that was chewed and the propeller
     on the plane that did the chewing. 91

36	Aircraft landed in the water off Kona.	92
37	Away to increase our earnings by $75,000 on the money the company already spent, fly straight from point a to point	93
38	Mechanic left the belly panel unfastened, pilot lost panel over Diamond Head.	95
39	Foiled a wheels-up landing Because a Mechanics were aboard	96
40	The aircraft lost two engine, company scrapped aircraft	97
41	The aircraft that was to come apart, would fly another day	98
42	At 5:00 AM the aircraft fine, at 17:00 an engine change	99
43	President disappointed, owners mortified.	100
44	Emergency landing on Molokai.	101
45	Aircraft ground looped in Kona, declared a total lose.	102
46	A drinking pilot is a dead pilot	104
47	The President and CP got fired six months after I left the Company.	105
48	Aircraft # 57 was ground loop	106
49	My Duties	108
50	Air Cargo Enterprise.	111
51	Scenic Air Tour (night shift)	114
52	Barbers Point Flying Club	116
53	Island Airlines	117
54	Pacific Air Express	118
55	Teaching Basic Electronic High School	120
56	Scenic Air Tour	121
57	AVMAT Teaching Aircraft Subject	123
58	Maintenance Department	125
59	Daily Flight Inspection	127
60	Starting Aircraft Engine	129
61	Read It And Believe It	130
62	Emergency Code	131
63	About The Author	133
64	Bibliograpy	134
SYNOPSIS		135
AUTHORS BIOGRAPHY		137

AIRCRAFT MECHANIC

A group of individuals with a common interest and back ground who join together to establish the highest possible standards in workmanship and to expand knowledge and practice in their chosen field.

SYNOPSIS

My NOVEL is about my experience while working as a aircraft mechanic at PANORAMA AIR TOUR. They had a fleet of 18 Beech 18 aircraft. Which they used to take the paying public for a five hour ride to see the State of Hawaii from the AIR. I never went on the tour myself. I don't like airplane but I worked on them and did a very fine job it seems even if I do say so myself and was always interested to learn about WHY AIRPLANES CRASH.

My company had their share of these problems along these lines. For the 1st year I worked there I did routine maintenance on the aircraft. The second year I was more involved in working on the engine and landing gear. The next 3 years I was the Lead Mechanic and the last 4 years I was the Director of Maintenance.

The Pulse of the company improved by 300 %. I had started a PREVENTED MAINTENANCE RROGRAM and the problems we had before just Disappeared.

This novel is based on fact, but I might have borrowed some things that I am not sure of so I'll say if is a novel of fiction based on fact.

When I started at PANORAMAR AIR TOUR I HAD 3 MECHANIC THAT HAD LICENSE AND 3 DID NOT. We had 18 BEACHCRAFT aircraft to maintain.

THE MORE YOU KNOW THE FURTHER YOU GO

FORWARD

General Aviation is a varied business at any AIR PORT in the country the watch word is SAFETY. In HAWAII we may be more watchful that the rest.

SOME OF THE ACTIVITIES ARE AS FOLLOWS.

1. Crop dusting
2. Air Ambulance
3. Newspaper delivery
4. Search and Rescue
5. Traffic monitoring
6. News Reporting
7. Police Surveillance
8. Public Health
9. Construction

These Activities and more like the Air Tour Business who make if possible for visitors to see the entire State of Hawaii with the All Day Eight Island Air Tour.

Unfortunately operation like this has accidents and here are listed of the 11,000 incidence and accidents that happen every year in this country.

July 24, in-engine Beech craft owned by Alii Air Hawaii Inc. and chartered by Panorama, crashed just after take off a few hundred yards off the end of the runway at Honolulu Airport. Five people were killed; four injured.

April 11, 1974 a crash on the slopes of Mauna Loa at 7500 feet killed all 10 people aboard.

March 3, 1976 aircraft veered off the runway at Kahalui Airport on Maui and ground looped it when touched the ground.

Airport on Maui and ground looped when it touched the ground.

The same day that same aircraft veered off the runway at Lihue Airport on Kauai, went through a fence, jumped an irrigation ditch, buzzed through a cane field and ended up in another irrigation ditch. No one was hurt but one landing gear was sheared off.

September 14, 1976 a belly landing in foam on the runway at Honolulu Airport after a landing gear of a twin-engine Beech craft could not be lowered. One of seven Japanese women tourist was treated for shock.

April 15, 1977 a landing on the nose of a twin-engine Beech craft with 11 people aboard, when the nose wheel collapsed at Lihue Airport on Kauai, no injuries.

September 20, 1979 a twin-engine Beech craft skidded down the runway at Keahole on the Big Island. After the right landing gear collapsed on touchdown there were no injuries.

June 22, 1980 a twin engine Beech craft 18 crashed at Honolulu Airport shortly after take off on a test run. The pilot who was the only one aboard was treated for bruises.

October 3, 1981 a twin engine Beech craft 18, went in to a steep bank and slammed into the waters 1/4 mile off Queens Beach.

April 8, 1983 a Beech craft 18, twin engine aircraft ditched in the ocean half a mile off Keahole Airport on the Big Island. One passenger nearly drowned. The pilot and eight passengers escaped serious injuries.

July 23, 1983 two Panorama Air Tours aircraft, each carrying ten passengers, collided on the runway of Kaului Airport on Maui. Two passengers were treated and released at Maui Memorial Hospital for elbow injuries.

March 29, 1984 a precautionary landing at Kaneohe Marine Corps Air Station to check an engine vibration. No injuries.

October 30, 1986 a forced landing on Halewill Road no injuries.

November 1987 Yesterday's crash landing in Kaplolani Park. One injured.

Flying the Hawaiian Skies. ALOHA!!!

THROUGHTS FROM THE PAST

The way a man dies is less important than how he lives.

The fool who keeps quite, will pass as a wise man.

You can fool some of the people all the time
All of the people some of the time
But you can't fool all the people all the time

A. LINCOLN

VOCABULARY

FLAPS—control surfaces used for slowing down, for landing.

PROPELLER—pull aircraft through the air.

RED LINE-indicates the limits high and low an item can withstand.

OIL PRESSURE—the oil pressure developed while the engine is running, recorded on a gauge.

DUZE FASTENER—metal fastener to hold panel etc. on air craft.

SHIMMY DAMPER—devise to dampen vibration.

GOVERNOR—controls propeller under normal conditions.

MANUAL LEANING—save fuel, by restriction the flow, when no altitude compensatory is part of the system.

FLIGHT SERVICE—federal agency to help pilots find their way home and other things.

EIGHT-ISLAND TOUR—tour the eight islands which make up the state of HAWAII.

MECHANIC'S CREED

Upon my Honor I swear that I shall hold sacred rights and privileges conferred upon me as a certified mechanic. Knowing full well that the safety and lives of others are depend end upon my skill and judgment. I shall never knowingly subject others to risks which I would not be willing to assume for my self, or for those dear to me.

In discharging this trust, I pledge myself never to undertake work or approve work which I feel to be beyond the limits of the knowledge, nor shall any non certificate superior to persuede me to approve aircraft or equipment as air worthy against by better judgment: nor shall I permit my judgment to be influence money or other personal gain, nor shall I pass as air worthy aircraft or equipment about I am in doubt, either as a result of others who have worked on it to accomplish their work satisfactorily.

I realize the grave responsibility which is mine as a certified airman, to exercise my judgment on the air worth-in Esc of aircraft and equipment I, therefore, pledge unyielding adherence to these precepts for the advancement of aviation and the dignity of my vocation.

AUTHOR UNKNOWN

PLANE TALK

It all began in 9/3/57 when I entered Northrop Aeronautical Institute Inglewood, California. Airframe & Powerplant Mech. My grades are as follower

Subject	HOURS	Gr.
Basic	140	84
Electrical	140	85
Woodwork	70	84
Welding	70	78
Sheet Metal	140	87
Engine A—Light	140	86
Induction System	140	83
Engine B—Heavy	140	80
Hydraulics	70	76
Propellers	70	85
Finishing	70	85
Rigging	70	81
Radio E & Ins	70	85
Auxiliary Cyst	70	80
Eng. Insng. Inset.	140	88
Repair Station	210	84

Graduated 9/12/58
Entered 12/ 15/ 58

JET ENGINE OVERHAUL
AND MAINTENANCE

My grades as follows:

	Hours	Grades
Phase #1 Development of Jet Engines, laws of physics & nomenclature	70	86
Phase #2 Thermodynamics, Combustion processes & types of Jet Engines	70	86
Phase #3 Engine Performance, Materials and Stresses	70	85
Phase #4 Jet fuels & lubricants, fuel and oil system	70	86
Phase #5 Electrical & Ignition System	70	85

Phase #6
Jet Engine Operation &   70   84
Testing

Graduated 12/9/58
Receiving my A & E License 1/9/59

I returned to Hawaii after I finished school and the first job I got working in my chosen field was with Murray Air working on their crop dusters. They had Sterman Biplanes powder with a R-985 engines and did they stink. The chemicals and Fertilizers they used were unfriendly toward Aircraft Mechanics.

I worked on both the airframe and the powder plant. The hardest thing about the airframe was replacing the bolts that were rusted away. They were attact by the chemicals and by the fertilizers and the bolts had to be replaced every three months or so.

If you mention stainless steel bolts and nuts to them they acted like they were hard of hearing but they acted like when you ask any question about any thing. I guess there was quite a difference in price between stainless steel and steel bolts.

I remember this one encounter I had with the Foreman of the shop. He came up to me and ask" it I had been assigned to anything this morning.

I said, NO.

He said, "follow me!"

So I follow him to the most distance place and still be in the shop.

He showed me 6 clinders and he told me.

"They were steel barrels I want you to take that electric drill motor and that cylinder hone. Put a .003 choke back in the barrels, all 6 of then, call me when they are ready."

I worked nights when I was going to school and I never slept in any on my engine classes. The only way I know of to repair those cylinder is by 1. exposing the steel to ammonia or cyanide gas while the steel is very hot.

The steel soaks up nitrogen from the hot gas which forms iron nitrides on the exposed surface. As a result of this process, the metal is said to be NITRIDED.

Number 2. b. porous chromium plated wall of cylinder barrels have been found to be satisfactory for practically all types of engines. Dense or smooth chromium plating with out roughened surfaces, on the other hand, has not been found to be generally satisfactory.

I never saw the Foreman for two day after that and I don't know what happened. The person in the engine overhaul shop took the 6 cylinder and put them back on the airplane.

The owner of the plane took his wife and another couple on a short ride. They took off and the engine was to weak to lift them off the ground. The owner was looking for the foreman, the foreman was looking for me, and I was in the engine overhaul shop doing what I like to do.

The foreman came up to me and said, "the 6 cylinders you were working on failed the test, they put them on the engine and they wouldn't even lift four passengers off the ground."

I just have this to say then it you want to fire me that's O.K. "First, For a foreman to give the instruction you gave me shows he does not know what he is talking about or is trying to hide his lack of knowledge or he is afraid of his job."

Second, I have been out of school about one year and we were taught there is only a couple a ways you can fix those cylinders by Nitride by the manufacture, or chrome plating done by several manufactures.

You could have ask me I would have told you what you obvious know by now, that s all I have to say I won't mention it again. I'll stay in the engine overhaul shop till you make up your mine about me.

I'll even go to the owner and explain it all to him if you would like me too.

"That's O.K. lets let it ride as is I'll get back to you in a few days."

In a few days he came to see me and said, "the owner has calmed down and he wants us to send the 6 cylinders out to have them chrome plated so lets just forget it happened at all. O.K."

I was working in the engine overhaul shop and we had just finish this R985 engine for the FAA. The company I worked for had an engine run-up stand but no means to connect the engine to the stand.

You would think a company of about 30 employees would have at least one competent welder but to no avail. As soon as they heard aircraft engine they hid there head in sand.

It I am not working at the airport I am employed at my Father in laws Steel Construction Business where I pick up several pointers on arc welding, like I need now.

I decided to do the welding my self and if it fall on the ground, ruins the engine I'll have something to chalk up to experience.

Looking around the scrap pile I found a piece of 4" angle iron 24" long. I cut it in 4" long I brought the engine out with the fork lift, put the engine against the stand where I wanted it.

Then I tacked the angel iron in place two pieces for each point the engine was resting. After they were tack on I removed the engine with the fork lift and set it back out of the way until the welding was finished. The welded two pieces on the left and right and another two at the bottom of the Engine. Like a triangle when you stood back and looked at it.

After completing the welding I took the ½" drill and drilled a 3/8 inch hole in each of the three engine mounts I had just welded on. The holes were just out side of center toward the front.

I put the engine back up there and drove three 3/8 "bolts through the engine mount and removed the fork lift it did not fall, "YET SOME ONE SAID,"

I put a nut on the three bolts sticking out of the engine mount and started to hook up the engine controls, the club prop on the engine and I was ready to start it up.

I got all the controls set, the mag. SW was ON I swing the prop the first time nothing happened, I tried one more time, the engine back fired, I tried one more time and the engine was RUNNING, I brought the engine idle up to 1000 R. P. M. and looked up and half of the day crew was out by the engine clapping I took a bow and thank them.

As the day wore on I increased the RPM every 30 minutes and it look like I would get by this time. About 3 PM I shut the engine down, removed the club and disconnected the rest of the engine lines.

I took the engine down with the fork life and set it in the shop on the engine stand. Then I proceeded to cut the engine mounts off the engine stand and put the welding machine and the rest of the tools away and went in the shop to clean up its about time to go home.

A couple of days later the big Boss came to see me. He said, since I did such a good job on that engine that he was prepared to give me a nickel raise.

I told him the company needs that nickel more than I do and knowing that I would not except it, he had nothing else to say.

All the time I was working at Murray Air I was going across the runway to see if PAN AMERICAN AIRLINES had an opening for a A&E Machanic and one day they did.

I told the lady I wanted to apply for that A&E position she had listed on the board. So she had me fill out the application and go to the Doctor for my physical examination the only thing wrong I was 10 lb. over weight so I went to work.

The days went by I spent every minute in the company library studying the system of the 707 Jet Liner. After the third day they called me in the office to tell me (they had to let me go) because my eye sight was short of the required (depth perception according to the Company rules)

How he put it, 'if I was setting on a tug pulling a 707 I couldn't be sure the wing tip would miss an object as the plane went by it' I've watch their planes being towed they always had a wing walker watch out for such objects.

Later I found out why they let me go it, seems PAN AM at this time had a rush of over weight people complaining of back troubles and PAN AM wanted it to stop.

The chief mechanic told me I was the best prospect he had seen in twenty years but he couldn't change their mind. I got my tool box and said my good-byes I went to Electronic class three times a week. A person who was on my crew at PAN AM asked me where I have been they didn't see me for a couple of days.

I told him my sad story he said,

"I should have went to the UNION office, once they hire you, they can't fire you until they talk to the UNION rep."

I told him I have sighed the paper. Thanks a lot for your help and concern.

I finished night school and got a job repairing Radio and TV for a shop in town waiting for something to break at the airport.

In 1962 I started work at the Mailwell Envelope Co. running a three color press. Printing envelopes for the Saving Loans, Banks, individual Business etc.

This Company had a concract with the Army to do the maintenance on their Aircraft which included, the Beaver, Helicopters, Beech Aircraft F-8, etc. about 16 pieces of equipment in all.

The Captain wanted to fly this day so he chose one of the Helicopter and took off. A couple of minutes he came back and said,

"That thing shakes so bad I could hardly hold on, you had better fix it."

Me and a coworker had it towed in to the hangar and got busy Rigging the Helicopter. Rigging the helicopter coordinates the movements of the flight controls and establishes the relationship between the main rotor and its controls and between the tail rotor and its controls.

Rigging is not a difficult job, but it requires great precision and attention to detail. Strict adherence to rigging procedures is a must. Adjustments, clearances, and tolerances must be exact.3 We finished the rigging and at 3PM we called the Captain to try it again, he did, it didn't it was worse. They say, the Helicopter was designed after the Humming Bird, it should have been designed after the Eagle.

I was at work one morning looking for something to do. I spotted two pieces of engine cowling setting on the bench waiting for someone to come along and put them back shape. So I got my tool box and opened it up so I could get the tools I needed at the time I needed then.

I started working on the cowling one piece at a time drilling, riveting, pounding, straighten priming. Painting at long last the cowling was finished I stood back and though to my self, that's a fine job I did on the cowling.

The captain came by to look at them and said, who every done that job on the cowling, could work on my plane anytime. I think I'll do just that who every did that job on that cowling. Have him work on my plane every time it's in the hangar.

The Boss at PAGE Inc. left the cowling on the bench for a week, so every one can get a good look at it.

We done All Weather Radar Modification on USF and USD type aircraft held by the Army from 5/67 to 6/68. The contract was not renewed in 6/68 and I was looking for a job 7/68 and I found one 8/68.

I work for Channel Air Lift until 10/69 about 14 months was their Lead Mechanic what that means is, the FAA came looking for me incase there is a mechanical problem with our aircraft.

The Director of Maintenance had a bad habit of drinking Scotch all week end and leaving me with all the maintenance problems. This one week he drank and about Wednesday I went to the dock were his boat was tied up.

He had a 36' sail boat and did not know one thing about it. That is way he had it tied to the pier. I got on his boat and following the scotch trail. There he was stretch out on his hammock and could not get up (he had the gout bad.)

"The Doctor said, he will be in that position for a least three or four days."

I went back to work cussing the people for making him Director of Maintenance. I was right not expecting any help from him for the rest of the week.

We had two aircraft that flew every day. One went down for an unscheduled engine change when this happens I have to fly with the other plane and keep it in the Air until the other one is airborne.

I remember one of these times, we left Honolulu Airport at 12:30 AM. one morning. We were headed for Hilo, Hawaii with a load of bread on board. I don't pay much attention to the loaders but I notice that every pallet of bread was wrap in plastic.

Well the flight was going very smooth and we were at 7000 feet and we justed started to fly over the island of Hawaii. We ran into a rain storm it was really coming down I was sitting in the jump seat behind the Co-Pilot and look over my shoulder I could not see the bread it was raining as hard in side as it was out side.

The Cockpit had holes you could throw a bear through and the Pilot, Co-Pilot and my self was soaked wet, the reason for the plastic. The pilot called the tower at Hilo airport and requested to switch to instrument.

The tower gave the pilot a few instruction and said, "the runway should be insight very shortly."

It was, we landed.

Other times I went to Kauai to replace a starter on one of our aircraft, my cousin lived on Kauai and we would go fishing after I got our plane back in the Air.

The other time I had to travel to Hilo, Hawaii to check on the two engines monster, it turned out to be the fuel pump had to be replaced. So I made the necessary phones calls to get a fuel pump to me in 3or 4 hour.

So I started looking for a fork lift, the Engines on these aircraft is 12 to 14 ft. off the ground, C-46. A fork lift was almost a necessity to work on these engines away from the base. I found the fork lift an bent over it to get the pallet lined up and felt a pain in my back, coming down here on Hawaiian Airlines I felt a tingle in my legs.

I finished the fuel pump change and rode back on to Honolulu on our plane. I went to the company doctor as soon as I got back. After some test he told me I didn't have back problems I have the Gout. In possible GOUT is for KINGS, PRESIDENTS, GOVERNORS, and the JET SET, not an A & E mechanic who at any given time don't have two dimes to rub together, I was living the good life and didn't even know it.

One of our Pilots was a used car salesman besides flying for us. He would buy a can, fix it, so it will run long enough for him to dump it on some one who knows less than he does about cars.

One day he bought a 55 Ford Mustang that he likes he drove it about a year using aircraft fuel the same one we used in our C-46s it was purple, rated at 145 octane. Octane goes to 100 over that its performance rating he was using 145 octane fuel in this car, and was ready to pass it on to someone who would agree with him, that fuel will surly keep the valves clean.

The car salesman filled up the tank with the purple stuff, put a for sale sign on the car and wait. A few day later a Service Man from the base came to look at his car. The service man looked it over and said,

"I like it."

He wanted to take it on a road test, sure the salesman said, handed the man the key and got in the passengers seat. The man put the key in ignition and turned the key.

It turned over but would not start, he tryed it several times but it wouldn't start.

The salesman ask me would I look at his car, "the engine turns over but won't start."

I look the engine over, every thing look OK, I ask the Man to start the engine, as soon as the engine was turning. I knew what was the matter.

I told the salesman the engine valves are burnt up, you need a valve job, the engine has no compression, see how fast the engine spins, there's no compression. You can sit there for the next hundred years the engine won't start, you need a valve job, it you are lucky. If you get the valves fix DON'T use the purple stuff again, I know its free, but eventually you will have to pay. Get your fuel for your car where every one else does, at a GAS station.

I worked at AIR HAWAII for about two years. They had twin Otters with P6A-27 Engines and were quite easy to maintain. The only trouble we had with them, we were constantly adjusting the idle. One night we were working, there were about 6 employees, THE D of M, Radio man, 3 A & E mechanics one person who cleaned and washed the aircraft.

Me and another mech. was adjusting the idle, I was on the top of the wing adjusting the idle, he was in the Co-pilot seat handling the controls. We had a code set up with our flash lights, I put my flash beam in front of him, once for increase, two for slow.

So we started, and he increased the rpm, I felt my hand slip so I signaled him with two flashes to slow down but I felt more pressure from the engine, he was increasing the RPM.

The rpm was quite high and I felt unstable hanging on to the wing, so I tryed the flash lights code again. I finally got his attention and he slow the engine down. I jumped down and went over to him, what are you try to do kill me I damn near got blown off that wing.

"He said, I am sorry I forgot the code."

I sent him up on the wing to adjust the RPM. I'll take over the controls.

1

Ma Worldly.

When I think of a witch I think of Ma Worldly She held the lease on the hangar in which the Air Tour Co. was house when I started working with when in 1971.

We had one space in the hangar to take care of maintenance problems as they arose. For example, weigh the aircraft every three years or sooner it the need arose. One hundred hour, and annual inspection etc.

In 1975 the Federal Aviation Agency disseminated to all Beech-18 owners a mandatory air directive (AD) note to install a stainless strap across the mainspar extending into each wing.

This would eliminate the need for X-RAY inspection of the wing and landing gear fittings. For the next three months our space in the hangar had a different aircraft in it every two weeks.

We had so much activity going on that we wore the paint off the floor. Mrs. Wordly from that time on she constantly harnessed us about the lest little thing.

For example, I was out by the gas pumps (she sold gas, oil and parts to small Plane owners who rented space in her hangar or anyone else who was willing to pay the price) and she came walking up to me and said,

"You must have done it you look the type."

Done what I asked,

"Broke my phone so it won't rang in the hangar."

I know nothing about phones. I guess you must look eleswhere for the quity party. She insisted I did it and I had to threaten her with a law suit before she would loosen up.

She finally got the phone co. to look at her phone line. The phone Repairman told her a rat had gnawed the wire into and that's why her phone didn't rang in the hangar.

While we were doing the modifications on the Beech-18, she placed a garden hose, thirty five ft. in diameter, around the Aircraft we were working on. Anytime some shavings got outside the circle, we had to stop what we were doing and clean it up. She was a major thorn in our side, untill we move into our own hangar across the runway in 1976.

Ma Worldly was a pilot in her younger day. On Dec. 7, 1941 at 3000 ft, she encountered the Japanese planes on the way to bomb Pearl Harbor.

. . .

Facing the runway on the left side was a small Air Cargo Business who generally had one Beech-18 flying at a time. He had three sons, two of them got killed flying airplanes and they killed other people too.

The people who owned the business was always at odds with the FAA and was never up to date with their paperwork. They did jobs for different people and they were alway unhappy about their results.

The Police Dept. had a helicopter stored in a T-hangar when they were not flying it. This morning they did there walk around, they got aboard and took off.

They got about 300 ft. in to their flight, when the rotor drive frozen up. What happened was the idler wheel that kept the belts tight froze up. The belts 5 of then got hot and the rotor stopped working and the helicopter crash on the tarmac close to where I was work. the rotor stopped working and the helicopter crash on.

I worked at night, this accident happened during the day. I don't think anyone got killed in the in the crash, but both spent some time in the Hospital.

Another accident that comes to mind, one day this lady came to rent an airplane to go flying for one hour. She done her required per flight inspection.

All seemed in order she got in the cockpit and taxied to the runway. She received clearance and took off, she got 100 ft. in the air, the plane nose dived into the ground and kill her instantly.

The investigation revealed that (she took off with the flaps full down) Pilot error. There approximately 11,000 incidents, accidents in this country every year. ACCIDENTS PEOPLE DIE, INCIDENT THEY JUST GET HURT.

. . .

On the right, looking toward the runway was the sister hangar of the one we were housed in. During one of the x-ray inspection we did on our plane every 25, 50, 75, 100 hours of flight time.

We found a crack by the Wing fitting and had to take that plane out of service for its repair. We removed the wing and called the Beechcraft welder from Witchita Kansas.

We complete the job in the three days, the plane was back in the fourth day. Another thing that comes to mine, that happened in that hangar. One day I came a little before 5PM and I tuned to the station that broadcast the traffic report. A few minutes later traffic reports came in over the radio. I was looking out the shop window, there was the helicopter with the engine running, the rotor spinning and the pilot was standing on the ground next to his chopper, giving the traffic report.

On the traffic conditions on the PALI HWY, H-1, H-2, and the Leewward Traffic FLOW and the Traffic flow in East Honolulu.

Do you get what you pay for, or WHAT!!!

2

Chief Pilot got Fired, President got hired.

One Saturday I was working on the Aircraft brakes I went into the office. As I went through the door from the hangar, the owner of the Company came through the front door.

The manager of our Company was about to go out the same door I came in, when the owner ask him where he was going.

Flying, he said, the owner said, he could keep on going and don't brother to come back.

A couple of days later, I was introduced to the President of our Company she had several business in Waikiki and was very well connected to the hotel business for potential clients for our 8 island air tour business.

Every year after, the Company purchased two aircraft and hired two pilots and one mechanic to fly and maintain them. When I left the Company eight years later, we had 18, Beech-18 Air Craft and a Company that was worth five million dollars.

The writing was on the wall, the Company started to go down hill and it became like a snow ball nothing could stop it. Some people would work hard to build the Company up and other in authority position work as hard to bring it down.

There were I think 5 other Air Tour Business during my eight year's at Panorama Air Tour. Sometimes we would lease for the day a plane and a pilot for a one day tour. I know for a fact this one Companies plane's were wore out.

They called me one day to look at the landing gear on of one of their airplanes, the landing gear were completely wore out. The mechanism in such bad shape it could hardly lift the gear while the plane was on jacks in the hangar with the doors closed.

Some Air Tour Co, had aircraft that incorporate the 8 cylinder Ryley Conversion and there had trouble with the magnetos on the engine. Then some Company shouldn't have been in the Air Tour Business at all.

This one Company I work for for about a year and a half had English built air craft, they had every thick wings and rode like you were setting on a powder puff, very smooth.

They had two flayes from a maintenance point of view, first, they had compressed air instead of hydraulics to operate the landing gear etc. They leaked and gave no indication of doing so.

Secondly, there magneto's were very weak, the manual said, 60 flight hours was the most you could expect from a set of two magnetos and two distributors we were on the road changing magentas at least three day out of the week when we should have been back at the hangar over hauling engine.

Finally, the Forman decided enough is enough, he designed a modification that fitted two American made magnetos to the engine we never had that problem again.

The other problem we had the owner would let us fix this one because we had to throw away the lock nuts used to lock down the adjustment screw, when you adjusted the valves on the engines.

In the Overhaul manual it says, it the nut will go all the way down on the adjusting screw with out the need of a wrench throw it away and get a new one, the key word here is new, he didn't want to spend the money (he should not have been in the Air Tour Business either).

I was hired at this Company to Overhaul Engines. The Gypsy Queen 30, and Gypsy Queen 60. The smaller engine was for the 4 engine plane and the larger engine was for the 2 engine plane.

There was three people working in the overhaul shop at this time, the Forman, Charlie, and myself. Myself was putting the crank shaft and connecting rods together starting an engine build up.

I usually started with #6 rod and work through the engine to #1, #5,#4,#3,#2, went well, they felt good and I was satisfied that they would go the required 1600 hours.

Number 1 it didn't feel right, if was loose, I tried filing it a little according to the manual, it wouldn't tighten up, I even cut a little off each end of the bearing of no avail, the only thing I could do was change the connecting rod.

I told the Foreman what I did and what I purpose. He said, you can not be serious, the old man wouldn't go for it, you know how thrifty he is(another one who shouldn't be in the Air Tour Business).

What do you purpose we do, I want to replace the rod if we can't do that, I'll junk the whole engine because of one rod and start the tear down of the next engine and hope that all 6 connecting rod are good.

The Foreman said, let me see the rod that you don't like. I showed him the crank shaft and the rod fitted very well until he saw the #1 rod the color drained from his face he said, go and finish the engine. Another one that shouldn't be in the Air Tour Business.

I took out my little black book and wrote every thing that had transpires about the connecting Rod and ask the Foreman to sign it, he refused, so I ask Charlie to be a witness because I knew this engine wasn't going the required 1600 hours.

I finish the engine, we push it out in the line for the night crew to use it when its turn came up and come up it did. They put it on the left side of the aircraft and it perform very well for 3 1/2 months. Then one day on Tour between Molokai and Lani island the rod went right through the wall of the engine case and spilled hot oil on the red hot exhaust pipe and it all caught fire.

They had to declare an emergency and land on Lani because they were closer to Lani than to Molokai when the connecting rod went through the wall of the engine.

That was first fire emergency invoving an aircraft they HAD ON Lani for 25 years. The night crew went to Lani to change the engine and flew the plane back a couple days later.

In the mean time back at the ranch the FAA wanted to talk to the mechanic that overhauled that engine that had the fire and declared an emergency on LANI.

So I went to my locked tool box and got my little black book and started for the office. The foreman ask, where are you off too, I told him the FAA wanted to see me in the office.

He told me to stay in the shop and continue working on the next engine overhaul. I told him if I hear anything derogatory about my ability to overhaul these engines, I would go to the FAA with my little black book, I never heard a word.

3

The Company hires a Supervisor.

The Company was grow and they needed a supervisor, and I like every body else thought I was qualified and that I would get the job. The Director of maintenance occasionally called me in for consulting work for the repairing of aircraft engine.

However, I didn't get the job, the director of Maintenance thought I was to qualified (another person who shouldn't be in the Air Tour Business) did not hire me. He hired a fellow who did not have any experience working on Beech-18s I'll be damn it I was going to hold his hand.

He wore a neck tie to work (another person who shouldn't be in the Air Tour Business). The owner expected all people who was paid by him was working for him. To work and not ware a neck tie to work neck ties are for insurance salesman. The boss wants production, he wasn't getting it from the neck-tie kid.

About three months had passes, while drunk the supervisor all most ran over the owner with his car. He said, he didn't see him, that's quite true, when you are pie-eyed it's hard to tell a man from a fence post. I guess.

The owner was not happy about this, the next day he fired the guy with out cause he don't need a reason he is the boss. I got the job the next day.

At this time I called all of the maintenance people in I told them what I expected from each and every one of them and I read the riot act to each one and told them I they mess up it will be better for them it they walked the plank.

I was the maintenance supervisor for four years and during that time I no problem with any employee and the shop work went smooth. The director of maintenance would act up once in a while, I told the crew he goes home about 6:30 P.M. so just put off what you wanted to do untill then, out of sight out of mine.

Our Chief Pilot was a short man, I doubt if he reach 5 foot. One day he was flying our tour he wanted to lower the landing gear to land on the outer islands.

He put the gear handle in the down position and nothing happened. So the next thing he did was to crank the landing gear down manually but it wouldn't move.

So he got the gentleman who was setting in the CO-pilot seat to position the landing gear handle and hold it out against the spring (the spring on each side of the handle to hold it in neutral so it won't move while the gear is operating electrify) so he can jump on the handle and bring the plane safety home to all passengers delight, which he did.

4

Tank dry Over speed Engine change.

One morning, I was sent to Kona to repair one of our aircraft that had developed magneto trouble. After fixing the problem, the pilot and I got into the aircraft bound for Honolulu.

The weather was good and the flight was very smooth. I glance at the instrumental panel and saw the fuel warning light flicker. The pilot was looking out of his side window at the time so I knew he didn't see the light flicker.

I called his attention to it but he said, "don't worry about it. The right engines fuel warning light always blinks above 6,500 feet". About one minute and a half later, the right engine let out a roar that would awaken the dead. The right engine went from 1800 RPM to 2800 RPM inside of three seconds.

The engine run out of gas (that pilot shouldn't be in the Air Tour Service). Where I was sitting, you could hear the engine whining. The pilot was very busy for a minute or two, pulling the throttle back and switching the fuel selector to a tank that had fuel in it. Things quieted down after that and we flew on to Honolulu.

I told the Director of Maintenance what had transpired and that we should replaced the right engine before it flies again, because of over speeding.

He told me the governor is there to protect the engine and that it could not have happened the way I said, it did. (another person who shouldn't be in the Air Tour Business) In short, he did not believe me. (the governor will protect the engine under normal conditions but running out of gas IS NOT NORMAL.

The next day, that same Aircraft flew again. While flying down the canyon on Kauai, the number 8 cylinder blew off the power case, and pulled all ten of its hold down studs out of the case. Put a dent in the cowling above the number 8 cylinder.

Three days later, we traveled to Kauai to change the engine we should have changed back in Honolulu where the equipment we needed is available. It we had change the engine in Honolulu knowing that it did have a period of over speeding we could have save:

>Two pound trip ticket to Kauai
>Freight for the new engine and the old
>Over time pay for two A&E Mechanic
>Sleeping quarters for the two
>Meals about six

Just because the Director of Maintenance did not know the limits of the governor which does it's job under normal conditions. The last time I heard about the Director of Maintenance he was in Arizona overhauling golf carts.

5

Gear will not come down Electrically
that's what the clutch is for Emergency!!

I had just gotten to work and as I walked into the hangar, I was told I was wanted on the phone. It was Flight Service they wanted me to know one of our pilots, having trouble getting his landing gear down and locked.

I walked over to one of the aircraft that did not fly that day and got him on the radio.

"What difficulties are you having with the landing gear"?

"It won't extend Electrically".

"Push the landing gear motor circuit breaker, which is located under your left foot!

"I tried, but it won't move"

"Push the land gear clutch in and crank the gear down, crank it away from you".

"I don't have to used the clutch, it should come down when crank in the down direction.

"Who told you that you don't have to use the clutch?

"No one, but on the aircraft I was checked out in, the landing gear would operate without pushing on the clutch. That's very true, but YOU ARE NOT IN THAT AIRCRAFT now, are you"?.

"So push the clutch in and crank the gear down and locked and bring that #@!%›&*_+ø aircraft home.

That pilot should have been told that only two aircraft in the whole fleet of 18 Beech-18 had the modification that, permitted the gear to be crank down, without using the clutch.

The fault belong to the chief pilot and his assistant, they logged many hours for training these pilots after the training, you should expect a pilot to know what aircraft has the modifications and which ones don't.

The chief pilot and his assistant would have class for the pilots and logged it as flight time. His name resulted in the initial F. O. G. and I think he has been in a fog since he realized this.

6

Aircraft Engines quit over Ft. Kam, 4 people die.

It was a bright sunny day in the Islands. This day in particular, the Company chartered an aircraft from across the field to take a group of ten people on the eight-island tour.

The pilot taxied across the runway to pick his passengers up who were waiting at our office in front of the hangar. He got the passengers on board the aircraft, taxied out to the runway, received clearance I assumed and took off.

The aircraft got about 400 feet above Fort Kamehameha and BOTH engine quit. The aircraft turned to its left and crashed into four feet of water the waves and surf pounded it into pieces.

The bodies of the people were remove, they picked the wreck up with a army helicopter, it was held together by the control cables. They deposited it by the Company hangar and people came for two weeks to view the remains.

Four people died and six survived, one of the survivors sued the Company for 3 million dollars and won. The pilot was an excetive at a local Insurance Co. he flew on the week ends. You can make a living flying air tour planes full time, its not for part timers.

Investigation revealed the pilot was not flying his regular aircraft that fateful day. On the aircraft he normally flies, the fuel selector at the full carries 78 gallon of fuel. The Pilot without realizing what he was doing, automatically placed the fuel selector where he normally would at the 78 gal. Position.

However, in reality, he actually placed the fuel selectors at the 25 gallon tank position. The company that owned the aircraft did not use them tanks because of weight and balance considerations, they kept only 5 gallons or less in them tanks to keep down corrosion and to keep the tanks in working order.

The pilot put the fuel selector in the wrong position and simply did not have enough fuel to keep the aircraft airborne.

As soon as the cause of the crash was published we color coded all of our fuel selectors as follows, 78 gallon tank GREEN, 25 gallon tank YELLOW, tank off RED.

Its simple mistakes like this one that make selling the air tour business or any business that used small aircraft compared to large aircraft (747) that they are safe as any aircraft. It turns out that regardless of the size of the aircraft the same basic problem exist, that is to get back on the ground safety.

7

Aircraft came into high,
replaced right landing gear & right flap.

It was about 2:30 PM and the aircraft were landing on Kauai for the last leg of the eight-island-tour. This particular aircraft came in to high and landed about two-thirds of the way down the runway, causing the aircraft to run off the end.

The aircraft crossed the cane field, jump over an irrigation ditch, a crossed more cane field and came to rest striding a second ditch. (another one that shouldn't have been in the air tour business)

The pilot got all the passengers out of the aircraft and bussed them to an inter-island carried bound to Honolulu. At first glance, it did not look too bad. We replaced the right landing gear, right propeller. Then we rented a huge crane, the aircraft was lifted up out of the sugar cane field and placed on the road used by the trucks that hauled the cane to the mill.

We towed the aircraft to the airport, checked it over one last time and flew it back the Honolulu airport. We completed the repairs and had the aircraft back in service with in the week.

The pilot made a boo-boo thats why he ended up in the cane field. He said, at the most critical time on preparing to land. A passenger ask him a question and that took his attention away from the business at hand, LANDING THE AIRCRAF.

8

Aircraft has no oil pressure right engine on
take off from Kauai, Engine change.

It was a bright, sunny day on Kauai and Howard my co-worker and I was on Kauai changing that we should have change four days earlier. I looked up just as our aircraft were landing from their all day eight island tour. At 5:00 P.M., the passengers were returning from boat trip up the Waive River which was part of the tour.

The pilot proceeded to direct the passengers to their respective aircraft for the trip back to Honolulu. The pilots got all the passengers onboard, closed and locked the door and taxied to the runway.

They received there clearance from the tower and took off. I notice one of the aircraft made a right turn and came around to land, while the other aircraft proceeded to Honolulu.

The pilot came over to where we were working and told me, "just as I lifted off, the panel gage indicated no oil pressure on the right engine".

I took off the right engine cowling and checked the main oil screen, the main oil bearing was in the screen it looked like sand, no oil no bearing. I put the oil screen in the aircraft and locked the door. I then put the top cowling on to protect the engine from the weather. Now we had two engine to change.

I asked the pilot what he thought had happened. "there was a ten or twelve year old boy sitting in the co-pilot seat all day. The oil shut is an Electrically control valve, playing with the switch, he turned it off. There is two problems here that are worth mentioning.

First, the pilot should not let any one under 18 set in the co-pilot seat.

Secondly, anytime a controlling switch is exposed to the public view it should be safeties wire with copper break away wire, or it should have a cover with break away wire to prevent this kind of incident every happening again.

The mechanic who done or signed the 100 hr. inspection prior to the incident should have received a find from the FAA for his negligent. That's why the Aviation Industry has rules and regulations and inspection so these things don't happened.

9

We got past Kona we declared holiday.

It became an inter-company joke that if we got PAST Kona without any problems, the rest of the day is a cake walk. I left the company on 11/15/79 and about one year they had lost nine of the 18 airplane the first year I was gone.

The aircraft crashed into the ocean be cause of engine failure the pilot feather the wrong engine, run out of fuel, crashed do to pilot error, flew down the canyon and waited to long to pull up, and other causes that only God and his pilots know.

When I become the D. of M. a few years later I incorporated a series of Preventive Maintenance practices that almost over night made the Company a going concern but every thing that lives must die and the company did.

10

Using 207 gallon of fuel, instead of the normal
250 gallon of fuel to SAVE money.

It was about 5:30 PM and the aircraft were returning from the eight-island tour. The Chief Pilot had another brain storm and declared a manual leaning campaign the week before. (manual leaning: moving the mixture control in the retard position until the engine's rpm drops then push the mixture forward until the engine's rpm rises 25 rpm.)

The pilot of one of our aircraft came over to me and said, "you will be proud of me, I flew the tour on 207 gallon of fuel instead of the usual 250 gallon.

I asked the pilot to stick around that I wanted to show him something. I told the mechanic to do a compression test on this pilots aircraft now. In about twenty-five minutes later the results were as follows:

#1 Engine TSO 1539.4 (engine change 1600 hours)

cylinder #1 64/80
      3 61
      5 65
      7 65
      9 60
      2 60
      4 61
      6 67
      8 60

#2 Engine TSO 667.4

cylinder #1 60/80
        3 65
        5 66
        7 62
        9 60
        2 62
        4 60
        6 60
        8 60

The minimum compression that is serviceable is 60/80. The average compression for Engine #1 is about 60/80 and for Engine #2 about 60/80. Without manual leaning these engine would have compression reading of #1 60 to 70, #2 70 to 80. Twelve cylinders out of the 18 should be changed.

I found the pilot and showed him the results of manual leaning (BURNT VALVES) and what it can do to the engines. After I explained the results, he was not a happy camper. Knowing the truth takes the fun out of it.

This madness about Manual Leaning, every Chief Pilot has to try it, he thinks it will work for him when it did not work for others. In my 8 years with the Company I did not see anyone make manual leaning a means to save money.

They might have flown one all day tour, the engines pay the piper eventually with burnt valves and engine change.

The aircraft comes in about 5:45 PM the crew starts at 6:00 PM. About 9pm I was walking along the line,(3 line of 6 aircraft each) when the aircraft electrician stop me tell me that one of the older mechanic is telling the young mechanic they don't have to do the(POST FLIGHT INSPECTION AS YOU HAVE LAYED IT OUT FOR THEM.) I told him thanks and I didn't hear it from you.

When he told me an older mechanic I wasn't a bit surprise that someone would have something to say and I am glad I found about when I did so I can smash before it grows.

I found the party, and asked him about it, he said, sure I did, we should be doing other things instead of trying to fine loose hold nuts its a waste of time.

I said, as long as I call the shoots that is exactly what you will be doing. We haven't had an unscheduled engine change or cylinder change since I started this program and as long as I call the shoots it will continue.

11

Engine runs rough due to Manual Leaning,
save money spend it on repairs.

It was about 6:00 p.m. and all the aircraft were in. I went to the office to see
how many aircraft would be needed for the next days schedule.

We flew as many as 21 aircraft on the eight-island tour. One of the pilots
told me, "for some time, both of my engines have been running rough".

I took the cowling off both engines, #1 engine had a crack across the top
of the cylinder from spark plug hole to spark plug hole. The compression
test on the rest of the cylinders averaged 40/80 and 60/80 they should be
replaced. The cylinders look burnt and the engine didn't have a drop of
oil any where the engine as any self respecting engine would have. Manual
Leaning may sure of that.

I checked the exhaust stack it was thin and had a white chalky powder
inside the stack (this person shouldn't be in the air tour business) not ever
as a janitor.

The #2 engine like the #1 engine was bone dry no oil any where. It even
look hot, the compression test reviled two cylinder with zero compression
and the remaining average 45/80 and should be replaced.

The exhaust stack told the same story running to lean white chalky
powder inside the exhaust stack.

I knew the pilots were manually leaning the engines to SAVE fuel, and
destroy the engine in the process. I returned to the office to ask the pilot:
What in the hell was he trying to do, kill himself and his passengers? I told
him what I had found out and why his engines were running rough.

He told me it was has job to fly. I said, to fly YES but to abuse NO!! It's
down right criminal the way you people are abusing these Aircraft. That aircraft
stay on the groung tomorrow I will let you know when its ready to fly.

12

Found crack in the Wing Fitting during X-RAY INSPECTION.

We had to inspect the wing fittings and landing gear fittings every so many hours of flight time. (take x-ray pictures of the fittings, and then evaluate the film, looking for cracks etc. in the fittings).

In 1975, the FAA came up with this AD note (Air Directive) that all Beech-18 owners with 4000 hours total flight, had to place a wing doubter on each wind and a steel strap tying the wings together, across and attached to the main spar.

We did an inspection one evening and found several cracks in the structural part of the main spar. We took the left wing off and called the welder in from Beechcraft Corp. in Wichita, Kansas. He came over in a couple days and measured, cut, welded and was finished the job in 3 hours.

We replaced the wing and sent the aircraft on a test flight, the aircraft flew on tour the next day.

13

Mechanic sat on top of engine to look for magneto drop,
could have been killed.

It was near sundown, a mechanic and I was working on one our aircraft
whose engine had a magneto drop.(magneto would not allow the engine to
maintain less than 100 rpm drop when running the engine at 1800 with the
other magneto turned off. If it goes over 100 rpm it is considered excessive
and must be corrected.)

I told the mechanic that I was going to run the engine, so stay back from
the engine, especially the propeller. I got into the aircraft and proceeded to start
the engine. Just before I pushed the start button, I look out the right side and
saw the mechanic sittine on the oil tank behind the propeller, so he could watch
for the magento drop while the engine was running, thats what he said.

I got out of the aircraft, called him down from the top of the engine
and told him I would not fire him, but for his own safety and peace mine,
I think he should try a different occupation.

I did not see him for a couple days, I thought he had returned to
Australia, but he showed up one day telling me he was hired as an electrician
for the city and county of Honolulu.

13A

55 has a problem

The aircraft were coming in, it was 5:30 PM. and I helping the mechanic chalk the aircraft as they arrived. A mechanic told me the pilot in 55 wanted to see me.

I got over to his aircraft, the engines were still running, he motion for me to come aboard. When I got aboard, the smell of rubber, wire insulation burning was very strong.

I look over at the right engines AMP meter it was fine I look at the left engines AMP meter it read charging 20 amps at idle. I asked the pilot to in crease the rpm on the right engine to 1800 rpm, the AMP meter went up to 80 I told the pilot to turn the Right Generator switch off.

Then I asked the $64,000 question:

"Since you left Kauai, how long have you had the GEN. SW. on?"

"All the way, its been on!".

"Didn't you have a clue, something was wrong when you could smell the smoke, it burnt the box that hold the voltage regulator and all the wiring to a crisp? "I wondered were the smoke was coming from, I thought I would have it check when I got back.

"Well your back and by the looks of it, you have burnt the gen. reg. box and all the wiring that goes to make up the electric system for the right engine, it will take 3 day to FIX. You monitor the instruments, you saw the amps were excessive Off the SW.

"Do you have anything to say?

"I am sorry".

"Consider your self lucky to be alive and have some thing to be sorry about.

13B

The Fog Factor.

It was Sunday after noon and I arrive at work feeling no pains and ready to hit it again for five days before I get another week-end off. The week-end the fella who thought of it must have been a genius.

I went through the office like always and out the back door into our hangar. What this an aircraft in the hangar it looks like they changed the left engine but they forgot the prop, its over there on the bench, WHATS GOING ON??

The last time I check the board it had 700 hr. to go before a change was due. Today being Sunday I guess I'll have to wait until tomorrow when the fog rolls in to fine out about that aircraft in the hangar.

When I got to the fog was not in but the President was so I ask her. What with the aircraft that in the hangar, the engine change is 700 hour away and setting in the and it looks like someone changed the engine without consulting with me.

If they knew what they were doing it would be one thing but don't, they done every thing but the propeller, thats a little trick and they haven't learned the trick.

I tell you this, it your Chief Pilot started to change the engine but couldn't finish and didn't tell about it, hell will freeze over before I work on that engine.

The fog rolled in about an hour later and she told him what I said, he didn't come near me that aircraft set there for 5 days, he wouldn't admit he was wrong because he would be admitting I was right, I'll finish the aircraft.

I was almost ready to work I look up there was the President of the Company standing there and she said, I know you don't get along with fog to well but for my sake, won't you please fix I that airplane today. I told fog it he every pulls that stunt again he is fired.

Yes I'll finish that aircraft tonight, because I stand to see a qualified adult cry. It can fly tomorrow with the rest of the birds.

13C

To cold for Mainlanders in Hawaii.

It was late January or early February and for winter we had only 10 Beech craft in the air that day and the phone rang.

A pilot in MAUI is trying to light off the aircraft heater because him passengers are cold and they are from Canada. He wants me to give him the sequences that will get the heated going.

I told him I took all of that circuitry that I could fine out of that aircraft when we first got the aircraft a couple years ago.

Why??? You ask me why I'll tell you, so some hot shot pilot don't blow him self and his passengers up the moment he lights off the heater. Trying for that $100 tip a what.

13D

Can't fix the airplanes & 9th seat.

The Company bought two aircraft a year that why we ended up with eighteen. This one year the aircraft bought has 8 seat they wanted one more put in the cabin so all of our aircraft could haul ten passengers.

Summer was coming up my kids (2 SONS) and my wife her heart wasn't in it. Would go to the mainland and play at the National Parks. One year we were gone for five weeks, the President was fit to be tied when I walked in the off. I got it from three sides, there were four aircraft that have not move for four days, no one knows how to fix them.

I fine that hard to believe the fog can fix it 3/4 of the way, haven't been taught the (trick) is that it. I'll go talk to my boys and tell you in half hour what we can have for tomorrow.

I checked with the supervisor and they didn't have any problems except for one aircraft it had oil all over the front of the engine I told him to take the spark plugs out of the engine front roll and use that concerted light beam and look in there and tell me what our see.

Nothing put oil he said Take the last cylinder and replace the cylinder that leaks worst, I be in the office.

They would fly tomorrow I told with a smile. I told the supervisor he could do the Mod. and I'll do the 9th seat installation when I get back next months it the Company said no! they know that I'll take five weeks again.

It was a short vacation and I am back so now the seat. My first stop was the FAA field off to talk to the inspector who does not visit me so often he say he can not waste Government time, nothing is ever wrong.

O.K. MR. Mau you are a good men. I want to put a 9th seat in one of our aircraft so point me in the right direction. I can't help you with that, that an Engineering problem you must go to the Engineering office at the

Federal Building and see Mr. DAU he is the Aeronautical Engineer for the pacific.

The next day I went to see Mr. Dau and introduced my self and why I was there. That is very technical requires high math. Do you happened to have a degree in mathematics. No sir, I don't it you will point me in the right direction I am confident I'll will get the job done.

Well lets begin, you are dealing with (G) forces here, one (G) for an adult male it is 170 pounds, you need 9 (G) forces forward, 4 (G) forces aft, 1 1/2 (G) forces left & right.

That it in a nut shell, you have to document it and submit it to Mr. Mau for his approval, it won't be easy but it is possible. I thank Mr. Pau and went back to the hangar to begin my project.

I thought it I got by the biggest of 9 (G)'s the rest would be like a walk in the park. I got our worn out fork lift and drove it in the hangar, it will have to lift 9 x 170=1530 pounds if I figure how to do this.

I located a pallet and place four 55 gallons barrels of engine oil on it, they weighed about 350 pounds each so four of them weighs about 1400 pounds. I found some concrete weights use to hold aircraft down in a tidal wave warning, we do have them. I used four of these blocks that brought the total gross weight to about 1600 pounds.

I wrapped two cargo straps around the pallet and used the fork lift to hold them in place. The floor piece was already fitted in place, and the track for the seat were installed.

I them got an old engine mount and fastened the floor piece to it with several C-clamps and attached the engine mount to the fork lift in the same way.

I stood back and look at the fork lift, engine mount, wood piece from the floor, and 9th seat, it look like some thing out of the past I couldn't invision a future that would look like this.

My next step was to fasten the seat belt around the cargo straps and take the slack out of then by raising the fork lift so they become taunt and I did.

So! the time has come to try it out, if I succeed I will be glad, if I don't I will be sad. I got back on the fork lift and placed the lever in the (UP) position the engine was idling when things tighten up.

I gradually increase the speed of the engine and look down and the pallet just cleared the ground. I let the pallet go until it just touch the ground until the weight was all on the floor and shut the fork lift engine off. That was the easy part now I have to convince Mr. Mau that's the hard part.

The next morning I went to the FAA office to see Mr. Mau he was on the island of Kauai and wasn't expected back untill later that day. I left a message for him to call me it he got back before I went home from work.

About 4:30 PM that same day he called me and I told him I had gone to the Federal Building to see Mr. Pau and I was ready for the test for forward (G)'s 1530 pounds would you care to watch at 9:30 tomorrow at the hangar.

I will be there he said.

About 9:20 Mr. Mau was there and we started. He look it over and said, if it holds together you have your self a winner.

I started the fork lift and slowly took it so the pallet was 12" off the ground and I held it the for a couple minutes and let it down untill the weight is all on the ground.

After you have finish the other three test, write it up on a (337) form and bring it to the office so I can place my stamp of approval on it.

I did the other test, for the AFT test I had to turn the seat around 180 degrees and for left side test 90 degrees and for right side test 270 degrees.

Then it took one day for writing and sketch work and 5 days to complete the 9th seat project. After getting the inspectors approval stamp I took the piece of flooring with the seat attached and put it back in the AIRCRAFT for witch it was made.

14

Collector Ring Vs short stack exhaust system.

The collector Ring Vs Short stack exhaust systems. Of the two exhaust system mention the short stack system does more damage to the engine than does the collector ring system.

Collector Ring Vs Short Stack
Exhaust System

At 1850 rpm, piston speed is excessive. The vacuum, created by the action of the intake stroke, suck in oxygen (AIR) through the exhaust valves. Due to valve overlap, (exhaust gasses still leaving the cylinder while intake gasses, fuel and air mixture coming in and both exhaust and intake valves are open at the same time, which leans out the mixture, but not to a dangerous degree.(short stack system) (Engine run Lean by design)

Considering the collector ring exhaust system. We find that during engine operation, the collector ring exhaust system is full of exhaust gasses.

Although they are sucked into the cylinders through the exhaust valves, due to valve overlap, they are inert gasses and do not contribute to combustion, so the engine runs rich by design, when using the Collector Ring Exhaust System.

Our short stack equipped engines are like a cars with no muffler, run them long enough and the valves will worp and engine will lose power.

If we do not manually lean the engines, which have short stacks on them, the AIR) oxygen that is sucked into the cylinder through the exhaust valves, due to valve overlap, does not lean the mixture enough to cause problems.

The operation manual which I read stated no manual leaning below 5,000 feet. If we investigate, I think we would find at this time aircraft N658 had the collector ring type exhaust system as do N56, N57.

I have never seen an operational manual for our aircraft equipped with short stacks indicating that we can manually lean the engine below 5,000 feet or any other altitude for that matter.

15

Manual Leaning is it necessary.

Manual Leaning is an exacting operation. It cannot be performed perfunctory by every pilot following his/her own rules.

The successful use of Lean Cruising Mixture requires that the pilot have a through understanding of engine requirements under different power, temperature, conditions. The pilot also knows when the specific steps must be amended it operation factors are not standard.

For example, we must stop to consider that rpm, manifold pressure, carburetor heat, air temperature, spark advance, torque pressures, lead fouling detonation, preignition and engine conditions are essential part of the Manual Leaning procedure.

The engine on the Beech-18 (AN985-14B) are not equipped with a carburetor that has an altitude compensatory. Therefore, I believe it is unrealistic for anyone to manually lean these engines at any attitude that w we fly usually less than 5,000 feet.

Except, in the case of (24) Manual Leaning is necessary for safe smooth operation of that aircraft.

By not leaning at any altitude and by cooling off the engine for two minutes prior to shut down, I believe we can look forward to an operation which will be free of unscheduled maintenance problems, which are unnecessary, wasteful and damn inefficient.

Manual Leaning was done at the following times between 9/12/75 to 2/17/76 5 month period.

Date	A/C/TT	Position	# of Eng.	Eng. TT	Aircraft Damage
9/75	7959	# 1 cyl.	# 2 Eng.	792	N888M P/S
10/75	5784	# 2 cyl.	# 2 Eng.	1571	N8067 CY/rep.
10/75	9661	# 2 cyl.	# 1 Eng.	7710	N2802G REP, Eng.
10/75	8925	# 3 cyl.	# 1 Eng.		N4303Y REP/3C
11/75	7688	# 2 cyl.	# 2 Eng.	646	N1850 REP./Eng.
11/75	9020	# 8 cyl.	# 1 Eng.		N4303Y RPS
11/75	8062	# 5 cyl.	# 1 Eng.	1266	N888M REPL./ENG.
11/75	9047	# 8 cyl.	# 1 Eng.	943	N4303Y REP/CYL.
11/75	6070	# 5 cyl.	# 2 Eng.	1191	N24H Repl. Eng.
12/75	8796	# 4 cyl.	# 1 Eng.	1200	N215H REPL./ENG.
12/75	8802	# 7 cyl.	# 2 Eng.	1243	N215H REPL./ENG.
12/75	7857	# 9 cyl.	# 1 Eng.	999	N618 REPL. CYL.

12/21 75 change Power Settings, 2000-2800 2100-31" CL.

Date	A/C/TT	Position	# of Eng.	Eng. TT	Aircraft Damage
12/75	8924	# 2 cyl.	# 1 Eng.	1424	N2802G REPL.CYL.
1/76	9874	# 1 cyl.	# 1 Eng.	1064	N24H REPL. CYL.
1/76	8283	# 2 cyl.	# 2 Eng.	1109	N888M REPL./CLY.
2/76	9076	# 9 cyl.	# 1 Eng.	1217	N215H REPL./CYL.
2/76	9081	# 3 cyl.	# 2 Eng.	1263	N215H REPL.CYL.

4/6/78 Company Policy Regarding Manual Leaning.

Manual Leaning will only be accomplished for the purposes of preventing engine roughness caused by excessively rich mixture as N24H experiences.

The NA-R9-B carburetor is the prohibiting factor and is installed on all Company aircraft, fuel consumption should be between 240 to 250 gallon per 5 hour tour and will be screen daily. Failure to comply may result in engine failure or worse.

N24H used 250 gallon on a 4.8 hour tour, on 1/11/79, before and after that date, consumption dropped to 210 gallons per day which means that they were still manual leaning the engine after the Company had published their NO MANUAL LEANING POLICY.

All exhaust valves leaked in all cylinders indicated that the Engine are running too hot and that as long as the Company lets the pilots get away with it, the passengers are flirting death.

Prior to this latest leaning craze, an engine with 700 hours total time had an average compression ratio of 75/80. In the case of N24H both engines had an average of 45/80 it the aircraft would lose one of those engine there's little doubt where the aircraft would end up.

This is the reason I left the Company. Over the eight years I work there, I had the disagreement over Manual Leaning five different times. Every time the company would hire a new Chief Pilot we went through this, I finally told them if they are trying to see how close they can come with out actually killing some one they don't need me.

The Chief Pilot could not control the pilots, they were so interest in saving money, instruct the pilots to fly in a straight line,(in stead of flying the passengers all over the in hopes of receiving that elusive $100 tip at the end of the flight) would have netted the Company $5000 per aircraft with out spending more of the Company's money.

(The series on Manual Leaning, I was assisted by a gentleman who spent 30 years in the Air force and was a Flight Engineer during that time. He flew as a pilot for my Company and was concerned as I was, how the Company demonstrated no control over their pilots).

I recognized his contribution but I forgot his name it's been about 20 years now.

16

Chief Pilot gets a PAY RAISE, I want mine.

A lesson in office politics. I had some business in the office this particular day. I told the office girl I needed some information out of the personnel file, about one of our mechanic.

She said, "the file is open, that I could go and get what I needed. While going through the file I run across the Chief Pilot's pay record."

He had received a raise of $500., two weeks ago from $1500. to $2000., a month. We had been payed the same salary up to this point, this meant WAR.

I got the information I needed (and then some) thanked the girl for her kindness, back to the hangar I went.

I got a pay raise form and filled it out for myself, from $1500. to $2000. per month (I always felt, under paid doing this job. I felt $4000 was not to much) I back dated the form two week and submitted it to the president.

She said, "what in the meaning of this"? "A pay raise for me" "you did not tell me the Chief Pilot got a pay raise two weeks ago, I wasn't aware WE were keeping secrets".

"I don't have the authority to give such a raise, I must get an OK from the owners".

"That's fine, you have two days".

Two days later, she notified me that she could only give me $250.

The FAA in 1976 designated the Chief Pilot as Director of Operation the first thing I did was to tell Mr. FOG stay out of my way. I went from Director of Maintenance to Chief Mechanic in a heart beat.

Thing were running smooth until one day I went to Kona to check on one of our aircraft.

MR. FOG was the pilot I check the engine out and found # 1 connecting rod had broken in to just below the wrist pin hole and the piston was stuck in the top of the cylinder.

We had three choices,

1. Change the #1 (link rod) (from another master Rod.)
2. Change the engine where the aircraft sit and save the engine, repair it later back at the hangar.
3. Fly it back as it was, destroy the engine, maybe we could save the crank shaft.

ANSWER: he chose # 3 to flew it back and destroyed the engine—the wisdom of the FAA's decision to put people like him in charge of anything is beyond me.

17

Battery Fire Reported Fire, Replaced Battery.

The aircraft were all in or accounted for. We were assigning the work to make sure everything is covered. When someone came running in to the hangar, yelling that an aircraft is on fire, one of ours.

We ran out side to see smoke coming out of the left battery compartment of one of our aircraft. I told some one to call the fire Dept. In a little while the smoke cleared and we opened the battery compartment.

The plus side of the battery shorted out and burnt the positive terminal off. We changed the battery, I wish all of our problems were that simple I think they are but when you add the fog factor thats when things get hairy.

18

Gear problem, down and locks he thinks, indication light would not come on

I went to work early to day I usually start work about 2:pm. My wife works,(thats the price you pay living in Hawaii) our two boys are in school so I went back to school working on my Ph.D. in Education I'll get it some day. (I got it Dec. 1992)

I had some paper work to finish that I started the day before.

I had been here about thirty minutes and the phone ring.

It was Flight Service calling to say that one of our aircraft had trouble with its landing gear.

The landing gear extended normally but the pilot had an "unsafe condition" in the cockpit (the landing gear indicating DOWN LIGHT did not come on).

I told Flight Service to have the pilot fly back to Honolulu. We could foam the runway and if it collapse, it would not damage the aircraft as much. Also, to cause even less damage landing with the flaps UP.

The pilot agree and I got ready for the worst but always hoped for the best. I figured he would be landing in about one hour and fifteen minutes. One hour later the phone rang and the pilot said, he done something or nothing but have landed safety on Kauai and everything is fine.

19

Pilot left the battery on for 3 hours on Kauai lost power,
unlocked gear, gear collapsed on landing.

It's 2:00 PM on the Island of Kauai and the aircraft are landing. They are on the final leg of the all day eight island tour.

The pilots taxied up to the parking area and unloaded the passengers for the 2 hour boat ride up the Wailua River.

The pilots arrived back at the aircraft at 5:00 PM just as the passengers were returning from there boat ride. The pilots helped the passengers get aboard the aircraft, taxied out to the runway, received clearance and took off.

A minute into the flight, the electrical power went off of one the aircraft. (engines are equipped with magneto which are independent of the battery circuits of the aircraft). Upon his arrival on Kauai the pilot forgot to turn the battery switch OFF. (another one who shouldn't be in the Air Tour Business) After taking off, he put the landing gear handle in the 'UP' position. The landing gear started 'UP' then the power went off. The alternators (AC Generators) went off the line.

Instead of the pilot cranking the gear DOWN, making sure it was locked, he came around and landed. The main landing gear was OK because of the special high strength close tolerance 3/16 inch steel bolts that held things together.

THE NOSE LANDING GEAR COLLASPED AND THE AIRCRAFT SLID TO A STOP.

We took another nose gear to Kauai, installed it, flew the aircraft back to Honolulu, GEAR DOWN it took two weeks to repair.

This was typical pilots performance in a crises they couldn't think unless they were setting down. Probably due to military training or other factors not known. Most of them were retired Navy or Airforce.

20

Aircraft hit a radio guide wire at KOKO HEAD crater pilot and CO-pilot were killed.

This one is for young pilots and would be pilots. It had been stormy for a few days. The weather man saw no break in the weather for several days. Joe (fictions) anxious to fly. His brother told him to leave the aircraft on the ground Until the weather breaks. He was hot-shot kid and had all the answers.

You couldn't tell him anything. He would go down the runway with the landing gear lever in the 'UP' position, so when the wheels left the ground, the landing gear would go 'UP'.

One day he tried this and the aircraft hit an air pocket. The tips on bout props, were damaged and he almost bought the farm that day.

The storm was still going strong and Joe decided to take off anyway. Half way to Molokai, the clouds were so thick he had to turn back.

The wind forced him off course, he hit a radio guide wire near Koko Head Crater and crashed into the mountain, killing himself and his passengers.

21

Mechanic filled the engine with water, two bottom cylinder removed and water was expelled, ran engine for 10 minutes.

It was about 8:30 PM, the aircraft were all in and one of the aircraft N24Hwas park near the near the hangar. The left engine was blowing oil out the breather, it appeared to be plugged.

We took the engine cowling (engine cover) off and I ask the mechanic's helper to place the water hose in the breather line from FORE TO AFT the direction which he, as a sailor, should understand.

I asked him again if there was any questions, before he started. He answered in the negative, there was no question.

I went back to my paperwork. A bout half hour later, the lead mechanic said, "you will not like this".'

I went with him and what I saw, I could believe. Instead of feeding water from the firewall aft, so we could clean out the breather line, he had run the water forward in to the breather and filled the ENGINE full of

When I got there, water was running out of a hole in the propeller shaft. I told the lead mechanic that he would have to take as many bottom cylinders off as necessary to ensure all the water evaporated and the only way to be sure is to run the engine for 7 minutes after the cylinders were replaced and talk to that helper I think he is working for the competition.

22

Nose Wheel collapsed on landing
at Dillingham Field scrap aircraft, for spar parts.

One of our aircraft, coming back from Kauai, developed a rough running engine and made an emergency landing at Dillingham Field. The nose gear collapsed. The company decided to use that aircraft for spare parts. One less aircraft to work on, one less pilot and mechanic lose there jobs. So goes the Air Tour Business.

23

Aircraft landed on its nose wheel,
fuselage look like an accordion, took a week to repair.

The weather had changed overnight and the aircraft had to take off to the West instead of the East (during the month of October).

That evening, the weather had not changed and the aircraft were landing from East to West. I was on the North ramp this evening and was standing thirdly to sixty yards from the point of touch down.

A this one aircraft came in, I noticed the nose wheel was lower than the main landing wheels.

When he touched down, he bounced three times on the nose wheel before the main landing gear touched the ground.

I went back to the hangar and met the pilot, I asked him it he knew what he had just done, he said, NO! So I took him by the hand out to the aircraft. The aircraft forward of the wings looked like an accordion.

The next day, I was drilling rivets out of the nose section of the aircraft which looked like an accordion. The Chief Pilot wanted to know it I thought the mechanic had damaged the aircraft while he was towing it. I laughed in his face. It took two weeks to repair.

24

Shimmy Damper fails in Kona.

The phone rang and it was from a pilot who had just landed in Kona. His nose shimmy damper failed upon landing, it shook so badly that he could hardly hold onto the aircraft.

I arrived in Kona and proceeded to change the shimmy damp I bled the system three times to save time (bleeding the system is the act of removing air from the system) and told the pilot we were ready for a high speed taxi (take aircraft to take off speed without taking off) the high speed was a success and we asked for clearance for a short test flight.

When the aircraft reached about 65 knots the front of the aircraft was vibrating two feet up and down. We got off the runway and proceeded to bled the system three more times.

We taxied to the terminal and told the passengers it the aircraft stilled shimmied we would put them all on the inter island carrier.

They got aboard and they took off it must have work. I had one hour before my flight to Honolulu, so I went looking for a shade tree and a cool one.

25

Mixture control rod came off, not safety.

Here we go again, Kona bound. When the aircraft landed the pilot could not shut the engine down with the mixture con control lever. He could shut the engine down with the magneto switch.(it is correct this condition before the next flight because vibration from the engine could close the mixture control the pilot would have no means to restart the engine.

I got to Kona and took off the bottom cowling on the left engine. A blind man could see what had happened, a mechanic night before, did not install a cotter pin in the hole provided allowing the nut to back off of the bolt, assisted by the vibrations from the engine.

I'll have a few choice words to say to that gentleman this evening, that I promise you. I replaced the bolt and nut, placed a cotter pin in the hole provided, and sent the pilot and his passengers on their way. I had an hour before my flight back to Honolulu, so I look for a shade tree and a cool one.

26

A. C. E. cargo aircraft (DC-4) lost an Engine on take off.

I was setting just inside the hangar talking to the lead mechanic. I looked up and saw Air Cargo Enterprises aircraft a (DC-4) take off and was about 500 feet above the runway. A few seconds later, I look back because I heard a loud explosion coming from the direction of the DC-4.

A few seconds later the #1 engine was rolling down the runway. In the meantime, the aircraft circle to the right to land on the reef runway.

The reason for the lost engine, was that number 1 propeller threw a blade and it went straight up. With the blade missing, this threw the engine out of balance and shook the power case from the blower case. The 52 studs that held two cases together were pulled out of there mountings.

Some estimated the propeller blade went to 40,000 feet. It must have been 2 1/2 to 3 1/2 minutes later that the propeller blade landed about fifty feet from where I was sitting.

It dug a hole in the black top. spun around and cut a big gash in the gas truck tire, we had to change it. The blade spun around again and slid across the ground, through the hangar, hit the wall, change direction, hit a 5 gallon can of oil and came to rest at that point.

Since I knew where the prop blade came from I loaded it on the tug to take it back to the owners who would be out looking for it very not to mention the FAA.

As I was driving along the taxi way the DC-4 that lost the propeller blade was taxing back to the hangar. The mechanic who worked at ACE was walking toward me and I told here in your propeller blade and here comes the aircraft that lost it.

27

Propeller would not go through a complete cycle,
link rod pin moved out of position.

I got a phone call from a pilot one morning. "I can't get my propeller to complete a full turn, it stops at # 8 cylinder" "Will it pass # 8 cylinder if you turn it in the opposite direction?.

"Same thing!.

I caught the next flight to Kona I remover the # 8 cylinder and could see that the # 7 link rod had broken its safety wire, and had slid down into the path of #8 link rod, (Master Rod is located at cylinder #5 and the link rods make up the other eight cylinder position) preventing the engine from completing a full cycle.

I pushed the link rod pin in place and safety wired it to the hole provided on the master rod. I replaced the # 8 cylinder and look for a shade tree and a cool one before I return to Honolulu.

28

24 aircraft flew this day, 18 of ours, 6 of theirs.

On a busy day we flew all our Aircraft, 18 Beech-18s (made by Beech Craft), and had to lease aircraft to handle the overflow in its heyday.

This particular morning, we had twenty four aircraft in line to take off. It got so bad Flight Service called us and requested that we send the aircraft out in set of six every 15 minutes.

When an aircraft takes off, a waiting period of three minutes before the next one may go. At that rate, we would have the runway for 63 minutes, we must be considerate for the others users of the runways so it went.

29

Aircraft that would not fly.

It was 7:00 AM and the aircraft were taking off (quite a sight) for the one day eight-island-tour.

One of our aircraft took OFF but did not lift OFF. He went around and tryed again at 90 knots, the aircraft would not leave the groung.

The pilot taxied back to the hangar, got his passengers on board a different aircraft, down the runway he went, when the aircraft reached 65 knots, it reach for the sky and they were off.

Meanwhile back at the hangar, the Chief Pilot and his assistant was looking the aircraft over.

I had gone into the hangar for a latter I wanted to check the aircraft out and see why it did not want to fly this day. I got back to the aircraft and the Chief Pilot informed me he will fly this aircraft now, more of the fog factor.

Without looking over the aircraft he and his assistant taxi the aircraft to the runway, got clearance, took off. The aircraft that would not fly, wouldn't fly for the Chief Pilot either. At 65 knots the aircraft should have taken off, but at 90 knots, the aircraft refused. He taxied back to the end of the runway and took off again.

When the aircraft reach seventy five knots, the pilot yanked back on the yoke. The aircraft just leaped into the air. When the pilot reduced power, the aircraft would fall. (take off power can be maintain for 60 seconds only, both of the engines should have been change on that aircraft due over speeding, the flight lasted 15 minutes.) He almost lost it twice, he had a near miss flying over the gasoline and oil storage tanks and once over the navy golf course. He just missed a tree.

They got back on the ground and taxied to the hangar. They started to look for the missing piece and found the battery COVER missing on the left side right under the pilots nose.

I asked the Chief Pilot why they didn't find the battery cover missing on the initial inspection that morning. The only explanation is that they haven't done it yet and its to late to do the inspection.

30

First wheels up landing the Director Maintenance surprise.

It was near 5:30 PM, the aircraft were coming in, all 18 of them. The phone rang, it was Flight Service. One of our aircraft could not get its landing gear down and locked.

The pilot wanted to know it we had any suggestions. The Director of Maintenance contacted the pilot on the radio of another aircraft that came in early. They talked the situation over and decided to come in with wheels-up.

I started getting ready and the first thing I did was to call Flight Service about foaming the runway. Flight Service wanted to know how much foam in feet we needed. I don't know as I never had an aircraft land Wheels-up before. They recommended 3000 feet of foam and I said, OK!

The foam was on the runway and everything was ready. The Japanese passengers were quite shaken up. Time was running out, fuel was low and at this time the pilot radioed that they were coming in.

It was dark by then and we didn't see the aircraft until he was about ready to touchdown. He cut throttles about three feet above the runway.

The aircraft slid about 1500 feet before it came to a complete stop. (just a little to much foam—we'll know better next time, we did not have long to wait).

We hurried over to the aircraft, open the door and helped the passengers to the waiting room in side the fence. Another group who did not waste any time getting out to the aircraft was L.C. and her channel 9 news hounds.

They asked me what happened. I told then I couldn't say, since I wasn't there when it happened, I am not a TV news reporter. The FAA inspector-in charge was standing close by so ask him. They ask if I had something to hide. I told then if I told them my version of what happened I would not recognized it as the same story after they edited it for TV.

My crew went back to the aircraft and got the landing gear down and locked. We towed the aircraft next to the hangar door and started the repairs that same evening.

31

Second Wheels up landing, he got FIRED.

We were working on aircraft (53) for the last two weeks, which had experience the Wheel-Up landing and the phone rang. It was Flight Service calling to let us know we have another aircraft flying around in the clouds with the same problem that (53) had but it was (55) this time and was there any special instructions for the pilot.

The pilot told Flight Service that he had tried everything possible and nothing helped so he was ready to bring the aircraft in Wheels up, DE JA VU.

Since I was on the phone with Flight Service I ask them to lay 1500 feet of foam on the runway to deaden the sparks that may appear when the aircraft is sliding down the runway during Wheels-Up landing.

The foam was on the runway and the pilot made a perfect landing at the edge of the foam. The aircraft slid 1490 feet, which was 10 feet on foam to much, I'll make the correction next time.

We accompanied the passengers to the waiting room and put then on our bus that was bound for Waikiki.

Knowing the routine, because we did it before, we towed the aircraft next to the other one that we were still working on. Now we had two aircraft to repair—so goes the air tour business.

The next day the BOSS flew in from San Diego and wanted to know "WHAT IN THE HELL IS GOING ON HERE, TWO WHEELS-UP LANDINGS IN TWO WEEKS. AT THIS RATE WILL BE OUT OF BUSINESS IN 9 MONTHS.

The Director of Maintenance could not or would not give him a creditable answer, so he fired him on the spot. When I got to the next day, before I got out of my car, the boss offered me the position of Director of Maintenance.

The first thing I did was to have our welder cut the ends off the tow bar and throw it in the trash. Then go to Kilgo and buy 2 pieces of iron fence tubing,(one is a spare) 10 feet long and weld the ends from the old tow bar on it and make us a new tow bar.

I told the crew use only the small tug for towing the aircraft. The first one that bends our new tow bar, because he turned to sharp IS FIRED.

I told the boss if was part his fault that two aircraft, made wheels-up landings. The last time you were here. You bought the large tug and did not leaves any instructions on it use.

We have been operating four years with no problems with the gear because we were using the small tug when they made to sharp of a turn with it, the tug didn't have power to continue around so it stop.

The person driving the tug had to disconnect the tug, straighten the aircraft wheel, hook up the tug and continue on his way.

With the large tug it the driver turn to sharp he continued on and the tug had enough power to exceed the turn limits on the nose wheel, the result is two wheels up landing in as many weeks.

The reason why the aircraft did the wheels up landing was because the tug and the tow bar was too heavy for our aircraft number one. Number two, the tow bar I destroyed was made from a two inch galvanized pipe, not tubing. Number three, the nose wheel had turn limits on them and they were exceeded.

When the person towing the aircraft turned to sharply, it bent the shimmy damper shaft and prevented the nose wheel from straightening up with in ten degrees of true. If the nose wheel was more than 10% out of true, it went up in the nose wheel well crocked and got caught in that position due to the bent shimmy damper shaft. In the meantime, the main gear motor kept running, since the nose wheel could not reach the shut off switch, the motor kept on running and the landing gear mechanism was torn out.

When the pilot was ready to lower the landing gear to land in Honolulu airport, the landing gear was already died. The pilot had on other choice, but land wheels up, which they did. It took two weeks to get this aircraft (55) flying again, the pilot help there, its as if he slid the aircraft on the runway.

In the case of (53), the pilot had cut the engines throttles three feet above the runway, when the aircraft dropped, it broke some of the structional tubing on the main spar, the left engine had to be removed so we could repair all the broken tubing. he broken spar. It took five weeks to repair that aircraft.

32

Buzzing sound coming from the engine, BEES?.

The aircraft came home about 5:00 PM, one of the pilots came over to me and s "My left engine is making a buzzing sound". "I don't know I wish you would take a look".

True to our policy of checking out each complaint, we took the aircraft into the hangar to hunt for buzzing noise. After looking for 1/2 hour and not finding anything, we took the aircraft outside and run the engine for a few minutes, nothing. We scheduled the aircraft for the next days flight and did not hear about the buzzing noise again.

33

Aircraft ended up in the culvert on Maui, change prop, change cylinder

I got a call one day that one of our aircraft was having some trouble in Maui. I sent (nothing like doing it yourself) a mechanic to replace an exhaust pipe, which was broken off at the flange a common problem with Beech-18s using the short stack exhaust system.

After he had finished with the exhaust pipe the pilot continued on tour. The mechanic was waiting for the aircraft to pick him up for the ride back to Honolulu. By the time the aircraft arrived it was very dark.

The mechanic with his tools got aboard and they started toward the runway. After traveling a few yards, the pilot decided he needed to turn the aircraft taxi lights on, because the ground taxi lights were not on for some reason or another.

The pilot was fumbling around in the dark (he didn't have a flash light) looking for the light switch, ran the aircraft off the taxi way into the culvert which was about 5 feet deep.

The left main landing gear stopped about 4 ft. from the edge, so it was all right. The weight of the aircraft was supported by the right wing. The right main landing gear was extended all the way out and was about 2 ft. from touching the ground. The right wing had wrinkles in it and the leading edge was damaged.

The Company hired a crane and lifted the right side of the aircraft and when the nose wheel cleared the concrete drain pipe, about 15 of the many spectators standing around watching, pulled the aircraft back out of harms way.

The next day myself and two mechanics went back to Maui to repair the damages. We patch the right wing, did a propeller shaft run out on the right engine (due to sudden stoppage), change the right propeller, replaced cylinder on the right engine, a couple days later the pilot flew the aircraft home.

The aircraft was extensively damage and took two weeks to repair, cost placed at $5000. or more, all because the exhaust pipe needed replaced.

34

Number one piston did not move up or down,
Replace engine or replace link rod Chief Pilot knows.

The phone rung at 10:30 AM but I didn't thank it was any of our pilots, since they leave Kona before 10:30 AM. It was a pilot and he said,

"The engine makes a big noise when idling, it starts fine, runs OK noisy". It was near 11:00 AM, I couldn't not get to Kona until one o'clock, maybe 2. So I recommended to him to put his passengers on an inter-island carrier bound for Honolulu. He agreed.

I got to Kona and took a compression test of the engine all cylinders were normal except #1 it read zero. I removed #1 cylinder and the piston was stuck in the top of the cylinder. The #1 link rod had broken off just below the rist pin hole.

To save the engine from further damage, I told the Chief Pilot we should change the engine where the aircraft is parked. He said, he would fly the aircraft home and we could change the engine back at the hangar. (more of the fog factor)

I knew my number was coming up, I could not take this person for too long, he knew just enough to be dangerous.

About six months ago, we got rid of one character, the FAA has plagued us with another one. So I replaced the cylinder, put the spark plus back in the hole provided and the pilot taxied to the runway and took off.

I had one hour before my flight to Honolulu, I went looking for a shade tree and a cool one. (hope when you read my tales, you don't think I am an alcoholic).

The next day, we dissembled the engine and found that the internal damage was so great we had to junk all of it except the crank shaft.

35

One aircraft chew-up the other one, change elevator on the plane that was chewed and the propeller on the plane that did the chewing.

On the island of Kauai the sun was bright and hot. The aircraft were on their last leg of the eight-island tour. The aircraft were parked and the passengers were off on their 2 1/2 hour boat ride up the Waiua River.

The passengers were back and the pilots were helping them find their correct aircraft for the flight back to Honolulu.

The aircraft started to taxi toward the runway in a single file manner. One aircraft up front slowed down for some reason, the pilot directory behind him did not, his left propeller chewed up the tail section of the aircraft in front of him. Included in the tail section were the: ELEVATOR, LEFT RUDDER, LEFT VERTICLE STABILIZER. The aircraft that did the chewing just needed the propellers change.

36

Aircraft landed in the water off Kona.

The aircraft was on Kona on the eight-island tour. The pilot loaded his passengers and proceeded to taxi to the run way got clearance and took off.

A few minutes in to the flight the right engine quit. The #5 cylinder, the master cylinder was blown off its base and the other cylinders could not keep their elliptical path so the engine quit.

The pilot was very busy feathering that engine, BUT he feathered the left engine by mistake, with both engines out, the aircraft fell into the pacific ocean, the water was 25 ft. deep and 75 yards off shore.

It just happened that the State Fire Dept. crash and rescue team was having drills near the airport, saw what happened and came to the rescue of the passengers from the sunken aircraft.

Everyone got to the surface safely except for one person, who had a difficult time getting out of the aircraft. She sued the company for one and a half million dollars and won.

37

Away to increase our earnings by $75,000 on the money the
company already spent, fly straight from point a to point

All the work was done and I was sitting at my desk wishing the phone would
ring. It suddenly dawned on me that the Company had been founded on
a five-hour tour flight time for the eight island tour. That was over 5 years
ago, I wondered what a 4 1/2 hour tour which we could easily fly.

Will start with two newly rebuilt engine (no new ones available for the
past 25 years rebuilt that it) which at the time cost about $7500 per engine:

(1) 2 eng. @ $7500 = $15,000.00
     (1) 5-hour flight = 1600 div. by 5=320 flights
     320 @ $115.00 (cost of tour per person) =$36,000 @ aircraft
     = $662,400

(2) 4.5 hour flight 1600 div. by 4.5 = 356 flights
     356 @ $115.00 = $40,940 x 18 = $736,920
     $736,920-$662,400 = $74,520

Saving per aircraft $5025
No of aircraft 18
Total savings $90,450
Cost of Eng. $15,000
Savings = $74,520

The straighter route they fly, the more money they save for the
Company.

Then the Company could pay everyone a bonuses they have been promising for several months. They could do this because more money would be coming in for the same investment, just by flying from point A to point B in a straight line.

Most of the pilots were interested in their own little world and some flew as much as 5.5 to 6 hours a day's tour trying to get that illusive $100 tip at the end of the flight. I took the results of my trying to cut down on tour flight time to the President of the Company. She was delighted that she could give everyone a bonus without spending more of the companies money.

The Chief Pilot was lacking in a lot of things, the main thing he was lacking was leadership. The pilots would not listen to him and he knew it.

After the President left the Company and started her own air tour business (she started with 10 Beech-18) all of her flights were nearly 4 hours long, she knew a good thing when she saw it). By flying from point A to B in a straight line we could have saved time, equip. and MONEY.

38

Mechanic left the belly panel unfastened, pilot lost panel over Diamond Head.

A license A&E mechanic was working in the belly section and had finished the job, he had not finished buttoning up the belly panel. He had a duze fastener on each side of the belly panel and someone said, lets eat lunch. (10 PM)

He left the panel as it was and went to lunch. After lunch he forgot he had not finished fastening down the panel, and went to do something else.

The next day, the pilot lost the belly panel flying over Diamond Head on they're way to tour the islands on the all day island tour.

He said, they got a lot of noise from the wind and engines. He returned to the hangar for another panel and continued on tour. A few days later the mechanic received a warning letter from the FAA telling why that incident will not happen again.

39

Foiled a wheels-up landing Because a Mechanics were aboard

It was a bright, sunny afternoon, a call came in from Kauai. The aircraft had engine trouble and the pilot thought the engine needed a cylinder change.

Two pilots with a couple of mechanic (pilots were trainees) flew to Kauai to change the cylinder.

They were flying at 5,000 feet and was getting close to the island. The pilot made his routine radio call to the tower to get his heading, wind direction, speed and was clear to land. The pilot put the landing gear lever in the down position and nothing happened. He tryed again nothing. He decided he would have to land on Kauai "Wheels Up".

The mechanic were watching their every move and never heard either pilot mention using the manual emergency hand crank for letting the gear down.

"Why don't you try using the emergency system before you land Wheels Up on Kauai. We have trouble enough receiving help when we have to change a cylinder".

The pilot ask the mechanic it he would show him how to use the emergency system. The mechanic pushed in the clutch, crank the gear down and they landed on Kauai wheels down.

After they had changed the cylinder, they crank the gear up flew back to Honolulu where they repaired the landing gear for the next days flight.

40

The aircraft lost two engine, company scrapped aircraft

It was about 6:30 PM and all the aircraft were in, the crew was bringing one aircraft into the hangar to change a cylinder. The mechanic finish the cylinder change, prepared he aircraft for a test flight.

Early the next morning one of our regular pilots took the aircraft for a test flight. During lift off, something went wrong, the pilot had to feather the left engine.

The one they had change the cylinder on the night before. The pilot flew around the field when something else went wrong.

The aircraft dove nose first in between the two runways, it hit so hard, it knock both engines off the aircraft. They rolled about fifty feet down the runway and stopped.

When the cost of repairs is more than the aircraft is worth, determines where the repairs are complete or not.

You get the feeling every thing is getting out of hand, there is so many mistakes being made, the good experienced pilots are leaving the Company and I think it is due to the fact, the Chief Pilot is in complete charge, it you get a bad one the owner has a problem, that is why the company is up for SALE.

41

The aircraft that was to come apart, would fly another day

I just returned from Maui, the aircraft was in hangar the crew had the interior out of the aircraft.

"What's happening ".

"When I lifted off, a loud noise came out of the tail section".

Another pilot working for a different company told the pilot the tail was coming off, of course they believe him. I inspected the tail section completely, the hold down bolts for the tail section where there and could be counted no loose or sheared rivets and found on evidence of pending tail section failure.

The crew put the aircraft back together, sent the aircraft a short test flight for the pilots benefit, the test flight was OK, we put the aircraft on LINE for the next available tour.

42

At 5:00 AM the aircraft fine, at 17:00 an engine change

The pilot got to the hangar this particular morning at 5:00 AM. He checked his fuel and oil and completed his run-up. At 7:00 AM, his passengers arrived he directed them on board, taxied to the runway, received clearance and took off.

He landed on Maui to let some passengers off and picked up additional passengers, who was going to the island of Hawaii to visit PELE at the Volcano.

About 30 minutes later, he notice the right engine oil pressure was low (minimum oil pressure 60 PSI) and the cylinder head temp was raising.

Both gauges were not at RED LINE yet (red line indicates the maximum stress a unit can with stand). When the pilot arrived at Kona, he call me on phone. We discussed the problem and he decided to fly back to Maui and would me when he got there.

I was doing some paper work when the call came in. The pilot had taken off from Kona and was 10 minutes into the flight when the engine quit. He feathered it and continued on to Maui on one engine, about a 30 minute flight.

The next day I went to Maui to check out the aircraft. The propeller would not go though a complete revolution. The main thrust bearing had froze up. The crew came to Maui that afternoon to replace the engine.

43

President disappointed, owners mortified.

In the summer of 79, the owners offered to sell the Company to the President for one and a half million dollars, it was worth more, she accepted.

However, before they could get together to sign the agreement of sale, the owners had another offer of 5 million dollars, so the Presidents deal was put on the back burner.

The owners pursued the latter, but it fell through. They ended up selling the business to a pilot who had more problems than the little old women who lives in a shoe.

44

Emergency landing on Molokai.

The aircraft developed engine trouble after it had left Maui. The pilot decided to declare an emergency and land on Molokai. He over shot the runway and landed in the pineapple field. He wiped out a set of flaps and propellers. The crew went to Molokai to repair the aircraft and the pilot flew it home.

45

Aircraft ground looped in Kona, declared a total lose.

The pilots could not make up there minds about who was going to fly (57) today. The Chief Pilot was instructed to fly (57) himself, it they did not have a qualified pilot on board. This particular day, he did not want to fly that day so he talked a lesser skilled pilot into flying (57) (57 was a tail dragger it had a tail wheel instead of a nose wheel).

An hour and a quarter after he left Honolulu Airport, the pilot ground-loop the aircraft trying to land in Kona, the first stop on the island tour.

As the aircraft came in for the landing, the pilot over corrected for the wind just as the aircraft was about to touch down.

The aircraft spun around 270 degrees, slid off the runway into loose lava cinder, snapped off the right landing gear. The right wind hit the ground and caused considerable damage to it. It damage the main spar, the right propeller and broke the mountings off the right engine.

When I got there 3 hours later, the manager of the Kona airport was directing a huge forklift under the aircraft to lift it up high enough to drive a low-boy trailer under the aircraft. They were off the mark and dropped the aircraft several times before they got it right. As far as the Company was concern the aircraft was a complete lose.

When I was flying down to Kona that morning, I had a vision of repairing the aircraft and putting it back in service but after seeing the aircraft after they got through with it I declared it a total loss. They finally got the low-boy trailer under it and deposited I t about 30 yards east of the tower.

When the other aircraft would go by it, his passengers would ask the pilot questions about what happened to that aircraft.

So as not to advertise what company the aircraft belong. One of the pilots decided to paint over the Company's name.

I notified the insurance company and they settled the claimed, we bought the aircraft back from the insurance company for spar parts.

46

A drinking pilot is a dead pilot

After I left the Company, there was a shake-up, some people were fired and others were laid-off. The manger they ended up with was a pilot who had more problems with alcohol. One day, he decided to go flying and took two additional people with him (no fun in dying alone). They took off and flew around the area for awhile. He told his passengers that they were going to fly up-side down, so sit down and fasten your seat belts.

He tried several times to turn the aircraft over but couldn't. He should have stopped there, he tried one more time and the aircraft turn over and made a nose dive into the pacific ocean, they all died that day and still waiting in water 3000 ft. deep to be rescued.

47

The President and CP got fired six months after I left the Company.

The President and Chief Pilot were fired six months after I left the Company. The owner asked me to be patient that things were about to change, I though it was time to go and do something else. I left the Company November 15, 1979. ALOHA !!!

48

Aircraft # 57 was ground loop

The pilots could not make up their collective minds who was going to fly (57) today. The Chief pilot was instructed to fly (57) himself if they did not have a qualified pilot aboard.

That particular day, he did not want to fly, so he talk a less skilled pilot in to Flying # 57. (57) was a tail dragger it had a tail wheel, not nose wheel, there is a different.

An hour and a quarter later after he left Honolulu airport, the pilot-ground looped the aircraft trying to land at the KONA airport the first stop on the tour.

As the aircraft came in for the landing, the pilot over corrected for the landing, the pilot over-corrected for the wind just as the aircraft was about to touch down.

The aircraft spun around 270 degrees, slid of the runway into loose lava cinder, snapped off the right landing gear. The right wing hit the ground and caused considerable damage. It damage the right side on the main spar, the right propeller and broke the engine mount on the right engine.

When I arrived there about 3 hours after the fact. The manager of the KONA airport was directing a huge forklift under the aircraft. To lift it high enough to drive a low-boy trailer under the aircraft. It took several times before he got it right. A far as the Company was concerned that aircraft was a complete loss.

While I was flying down to Kona that morning. I had visions of repairing that aircraft I decided after I saw it, it was a complete loss. They finally got the low-boy trailer under it and deposited it 30 yards East of the tower. When the other aircraft would go by it the passengers would ask the pilot question about what happened to that airplane.

So as not to advertise what company that aircraft belonged one of our pilots decided to paint over the companies name. We notified the insurance company and they settled the claim the very next day. We bought the aircraft back from the insurance company for spar parts.

49

My Duties

LEAD MECHANIC:

1. Interpret pilots discrepancies and a plan of action to solve various problems as they arise.
2. Assign maintenance personnel to various tasks, dealing with routine and unscheduled maintenance problems.
3. Check progress of work being performed.
4. Answer questions and solve the problems mechanic encounter while on the job.
5. Lead repair crew for outer island rescue mission, Diagnose problems encountered by pilot on the other island.
6. Replace engine and perform repair operations as required to return aircraft to Honolulu from the outer islands.
7. Liaison between management and maintenance department to ensure adequate communications for the efficient, safe operation of our aircraft.
8. After required maintenance is accomplished, log books are signed and aircraft released for the next days flight.

Director of maintenance:

1. Directly responsible for the safety of 18, Beech-18 aircraft, schedule required routine maintenance per
2. Federal Air Regulations, Direct repair of all unscheduled maintenance which was minimal due to the PREVENTED MAINTENANCE INCORPORTED.

3. Conform to Manufactures Service Bulletins and the Federal Aviation Agency's air directives pertaining to Beech-18 Aircraft. Perform and approved major repairs and major alteration authorized by Inspection Authorization issued by the FAA.
4. Responsible for all aircraft log books entries, weight and balance and the air worthiness of all Company aircraft.
5. Hired mechanics and cleaning personnel.
6. Advise and collaborated with X-RAY technician as to the quality of X-RAY film and interpretation to determine if crack exist in the main spar of the Beech-18 aircraft. The X-RAY of the Beech-18 main spar is required by air worthiness directives issued by the FAA.
7. Responsible for supplies and spare parts. Collaborate with mainland repair stations repairing the accessories and engines. Maintained an adequate supply of parts and related materials necessary for the safe operation of our aircraft, 18 Beech-18 Aircraft.

OPERATED UNDER FAR 135 and 90.

6/19/79

Letter From My Employer

To Whom It May Concern:

This letter is to advise that I have known _____ our Director of Maintenance for 8 years. I have observed his performance of duty as a manager on a daily basis and find him to be tenacious in maintenance detail.

He evaluates all facets of the problem areas and has EST abolished excellent maintenance system and procedures which have resulted in achieving a 100% in commission rate of aircraft in support of the operational needs of our Company.

This achievement is of particular significance in view of our geographical location, the lead time required for spare parts procurement from various mainland sources, and an engine overhaul exchange program with a company in Los Angeles.

He personally supervises all maintenance activities, teaches and encourages the younger men, and inspires his associates to constantly strive for perfection in each task. He is patient,

understanding and devoted to professionalism in the business of
Aircraft Maintenance.

Based upon this knowledge, I would certainly recommend
_____ to any prospective employer. Please feel free to
contact me in the event that further information is desired.

Yours very truly

50

Air Cargo Enterprise.

DUTIES: AS AN (AIRCRAFT INSPECTOR)

1. Inspect engine sections for leaks of fuel and oil, hydra license, etc.
2. Inspect studs and nuts for proper torque.
3. Inspect internal engine, cylinders, compression test, etc.
4. Inspect engine mounts, engine controls. engine accessories for defects, and all system for improper installations, etc.

Operated under FAR 121

I work at ACE for about 2 1/2 months. My first pay check I got from them BOUNCED. After waiting for three or four days the bank excepted it and all was fine till the next time and it was a week away.

The system they had to keep the flight time legal was the index card method, they had to go through the whole card system all most every day. They would put a yellow tag on item that had at least 100 hr. left on it, at 50 hr. an orange tag, for less than hr. they would put a RED tag and prepare to change engine or instrument etc. at the proper time, when the time was expired on that item.

While at PAT I devised my own system, with 18 Beech craft it would be impossible for me or any one person to stay out of trouble with FAA for paper work violation as the former D of M demonstrated several times.

On a piece of plastic 1/4 x 4' x 6' I wrote the tail numbers of each aircraft starting with tail #50 to #67 down the left side of the plastics. I spaced them the best I could to give ample room for each number listed on the left side.

Across the top I wrote the name of the of each item that was time controlled such as: Engine, Propeller, Governor, Carburetor, Fuel pump, ELT emergency unit, Life vest, Fire Extinguisher, Engine Extinguisher.

On the Trio. Gear Aircraft the FAA came up with an (AD) note, that every landing had to be recorded and after 70 landing the nose gear required a DRI-Check inspection of the nose gear where the shimmy damper was mounted.

I placed the board in the traffic lane were anyone can see it. For the mechanic to look at and see what aircraft needed a 100 hr. or what aircraft was coming up for the Annual Inspection it was all there for any to see.

The FAA commented one day that even though I am in Kona they don't have to wait until I get back to show him the log books or any other thing all they have to do is look. The short time that I work for ACE they got there other DC-4 flying and visions of sugar plums dance in the heads. Now they could get back on there feet, make money, pay off the creditors and quite this talk about going under.

Business was good, the office phone rang more often and the company could see light at the other end of the tunnel. Then one of the CO-pilot left the company, they had to hire one right away, so they got this fella who did not have much DC-4 experience, he had plenty of experience in helicopters he'll do, they wanted to believe.

For a couple weeks they did fine. Then one fine starry light night they were bound for Maui. The flight was smooth and every thing seemed to be going OK.

They got close to Maui airport and the pilot said, gear down SO HE SAID, the CO-pilot heard it as flaps down as they approached the runway the aircraft gear not down, the warning light came on. The pilot and CO-pilot stared at each other for a split second and the pilot went for the gear down handle but it was to later.

The aircraft with its 10 ton of cargo slid down on the runway and the sparks were flying from every where. The aircraft slid several 100 feet and came to a stop with a groan.

Meantime, back at hangar the workers was getting the next load ready to go, the phone rang as phones will.

The pilot told them the unhappy news and activity stop around the hangar for the night.

The next day the boss came in laid all the cargo workers off and informed the mechanics that they would have the double as cargo handles till further notice.

The boss had blue prints for modifying a twin engine aircraft by extending the fuselage 16' and converting the piston engine to TUBRO PROP, putting a heavier landing gear under the aircraft and make a fortune in the process.

Selling it to the U.S. Government and to other Governments around the world. He also wanted to but TUBRO-PROPS on DC3 an make another fortune but some one in Canada had beaten him to that fortune. The company survived for two months more but finally had to draw the curtain.

51

Scenic Air Tour (night shift)

NIGHT SHIFT DISCRIPTION OF DUTIES

- A. Directly responsible for 3, Beech-18s.
- B. Scheduling required routing maintenance per FAA regulations.
- C. Direct repairs of unscheduled maintenance which was minimal due to the Preventive Maintenance Program that was incorporated.
- D. Conformed to manufacture's Service Bulletins and FAA air directives pertaining to Beech-18 aircraft.
- E. Performed and approved major repairs and major alterations authorized by the IA issued by the FAA after passing they test.
- F. Was responsible for all aircraft log book entries, weight and balance and the air worthiness of all Aircraft.
- G. Responsible for up keep of all supplies and spare parts.
- H. Collaborated with mainland repair station in the accessories engine.
- I. Maintained an adequate supply of spare parts and related material necessary for the safe operation on the aircraft.

OPERATED UNDER FAR 135—91.

While at SCENIC AIR TOUR we did not have a hangar. The routine maintenance was done on the tarmac in front of God and any body else who wanted to watch. Some times for an engine change we would take the aircraft to PAC next door to the hangar PAT use to occupied 5 years ago.

The fella who own PAC had trouble with tool lost every time he hired a new mechanic, he would have to buy a new set of tools. The fired mechanic would walk off with them. The supervisor at PAT had the same kind of trouble. One day I came to work and the President call me in to the office. She said, he was selling our parts to any on the ramp who had the money to pay.

I confronted him and told him if it is true he'll have to go, he got his tool box and left. Things were quite tame on this side of the north ramp not much happening.

3/27/88

To Whom It May Concern:

RE: JLD

It is my pleasure to recommend this gentleman to a position in Vocational Education. I have known him for 3 years in a student in the Trades and Industry Program at the University.

He did him internship at Honolulu Community College as an Aviation Mechanics instructor in the spring, 1980. His performance has been outstanding. Students and teachers have nothing but complimentary remarks about his teaching.

He impresses me by his professionalism, excellent rapport with people, his honesty and sincerity. I rate him a mong the top 10% of our T & I degree Graduates and recommend him to you without reservations.

X

52

Barbers Point Flying Club

Did general Aircraft Maintenance on 16 aircraft of different needs. The aircraft were owned by the members of the club. The club furnish the maintenance if the owners would let the fly club rent the aircraft to members of the armed forces for a cut of the pie. It did not last long.

53

Island Airlines

They had five BEECH-18 and a while later they got a large aircraft which suck the life out of every thing.

54

Pacific Air Express

I work about three months on the DC-4 this company had. They were from the other company I worked for four about 2 mos. They had the aircraft just inside the fence and we were all working hard to get the aircraft in the air. I had finished inspecting the left side of the aircraft and the crew was busy working off the discrepancy.

I was working on the right side of the aircraft on the main fuel tank. I stuck my head in and started inspecting the tank. I came across and off colored patch and started to see what I could see. I dug at it with my corrosion finding tool. I dug at it and dug right through the other wall. Now we had a hole that you could throw a bear through.

I got down on the ground and went to the other side where the D of M was working and told him he should stop what he is doing. Come look what I have found on the other side. I want back to my task and two hours later he came over to my side.

"What do you have to show me that could wait until tomorrow". "Just get up here and stick your head the tank and you tell me".

He stuck his head in the tank and then jumped to the ground said," I think you just put us out of business".

"Well should I continue with the inspection or not". "The old man is on the mainland, I'll notify him of our discovery, he'll make the decision whether we continue or close shop. You might as well go home and wait for my call.

About the fifth day I receive a letter from the company, my services wasn't needed any further and thanked me for the service rendered.

AN AIRCRAFT INSPECTOR CAN ONLY BE AS GOOD AS THE COMPANY WANTS HIM TO BE.

. . .

This aircraft was cruising from Kona to Honolulu and just as it passing Maui a section of the roof about 15' to 20' blew off. One flight attendant was suck out of the aircraft at 20,000 ft. nevered heard from her again.

The aircraft made and emergency land in Maui (it was in all the news papers and on the 6PM TV news) the hair line crack or cracks must have been there a long time. Wanting some aircraft inspector to spot them. I wonder why they didn't.

FROM:

1. Effective August 2, 1982 the PAE, Inc. aircraft inspection department will report the General Manager PAE, Inc.
2. Close coordination between the Director and Inspector is imperative and direct communication is encouraged.
3. In the event a decision can not be reached on the serviceability of any item, a final decision will be made by PAE material review board consisting of:

   Mr.—Mr.—
   and Mr.—. The review board will request assistance from
   the FAA and other sources if necessary.

Signed:

_____

55

Teaching Basic Electronic High School

Waipahu High School 11/81 to 6/82 Full Time
Waianae High School 10/82 to 6/83 Full Time
Mililani High School 10/83 to 12/83(Suffered Stroke)

First year: Introduction to Electricity and Electronics, Electron Theory and Current, Simple Electronic Circuits, Ohm's Law, Batteries for Electronics, Resistance and Resistors, DC Circuits, Voltage Dividers and potentiometers, Magnetism, D-C Meter Theory, Generator and Transformers.

Second year: Inductance, Capacitors, A-C Circuits, Electron Tube Diodes, D-C Power Supplies, Lab Projects, Basic Amplifiers Circuits, Signal Level and Decibels, Basic Audio Amplifiers, Cathode of Coupling, Cascaded Amplifier, Cause of Distortion in Amplifiers. Lab Project: Build AM radio from kit.

Third Year; Feedback in Audio Amplifiers, Grounded Grid and Cathode-Follower Amplifiers, Basic Television Principles and Service, Lab Projects: Students work on their own TV and/or Radio. Each student builds a AM-FM Radio from a kit.

56

Scenic Air Tour

Director of Maintenance, Repaired and Maintained aircraft within the rules and regulation of the FAA. Worked for SAT this time about 6 months. I did not have to worry about the Fog factor he was no where to be seen, someone told me he was working for the FAA as an flight inspector, BIRDS OF A FEATHER, STICK TOGETHER.

The Chief Pilot was about 35 I did not brother him he did not brother me. We had 10 Beech-18s and we flew the regular eight-island tour plus six other shorter tours.

The boss had a bad habit of hiring these mechanic from the Far East they don't speak any English. They can answer any question you put to them, it always (YES) I caught on to that quick.

They had quite a few problems with there aircraft. When I started they had 10, couple years they had 3, after 6 months they invited me out the story goes this way.

I started in OCT. and for Christmas and New YEARS they expected from me personnel to give each mechanic a bonus out of my own pocket, they said, its custom back home, I told them they had better get started and get back home that were charity begins.

Well a few days later I went to work a little early and Chief Pilot want to take me to lunch. I know from experience when the boss wants to take you to lunch its either to hire you or fire you, I had a job, so it is to fire me.

It seems the boys from the Far East told the boss that it I come to work the next day, they will all go him so what could she do.

About two months later I went to her office to see how they were making out, they were down to 6 aircraft the boys could not fixes them, two of the aircraft had cracked wing fittings she was sending them to the Philippines there is no restrictions on the aircraft over there.

I started to talk to her and notice a lot of barrels with use oil in them. The boys uses them to put the oil in when they change it every 100 hr. inspection.

That's not necessary, all the time I spent at PAT I change the oil once, that was the 60 wt. break in oil after I replaced that oil it wasn't change again until the engine was change 1500 hours later. I did the same thing both time I worked for you.

Look at it this way, after a 5 hour flight these R985 engines use from 1/2 to 1 1/2 gallons of oil for each flight. That from 3 1/2 gallon of oil on the low end, 10 1/2 gallon on the high end. It your engine used 3 1/2 gallon in 2 weeks you would have new oil every 2 weeks based on the 7 gallons that you start out with, so just add 1/2 gallon per day until you change the engine oil at 1600 hours.

With the other engine you put 1 1/2 gallons per 5 hr. flight per day. In one week you change your oil 1 3/4 times per week. Only for the initial oil change, you shouldn't change oil again until you change engine at 1500 hr. this for R-985 engine only.

In 1985 we decided to build a log cabin on a lot we brought in 1969. It is located on the Gallatin River about 42 miles north of Yellowstone Park on HWY. 191.

I was recovering from the stroke I pick-up in Dec. 83. It took about 1 1/2 years to finish and we moved to Montana about the middle of June 1987.

We stayed about a month and a half, went fishing every day, had trout for dinner about 5 day a week. Then one day phone (as phones do) A friend of mine offered me a job teaching at the AVMAT School so we left Montana and I started 8/87.

We go back every year, in fact, I just made our reservations last week for our trip the last week of June 1, 1990.

57

AVMAT Teaching Aircraft Subject

AVMAT (HONOLULU COMMUNITY COLLEGE)

AVIATION MAINTENANCE INSTRUCTOR FULL TIME

Honolulu Community College 8/87 to 5/88

AVMAT:

Avmat 30—Aircraft Woodwork, Aircraft Dope & Fabric, Aircraft Finishes, Aircraft Welding.

"41—Aircraft Electricity.

"43—Hydraulics, Landing Gear Units.

"47—Inspection, Troubleshooting and Repairs of Reciprocating ENGINES.

"48—Turbo Engines.

"34—Rigging and Assembly.

Instructed students in Airframe and Power plant curriculum subjects.
When the year was over, it was the last year for the old curriculum. The state had receive much pressure from the major airlines to up grade the curriculum. To make the graduate more compatible with their systems

and get away from the notion that General Aviation was the starting point for anyone wanting a career in the Aviation industry.

When I took the test to get my A & E License, there were 7 test that made up the Airframe proportion and 6 test that made up the Power plant proportion 13 different test in all.

Then there was the Oral Test that lasted 8 hours, then there was practical test that tested you on your abilities on Airframe systems and Power plant systems.

Today the A&E test given by the FAA or their agent has only three test, when I took the test it consisted of 13 tests.

General Aircraft Maintenance
Power plant Maintenance
Airframe Maintenance
The Oral test is done by the Designated Inspector.

The practical is done by a private individual who is the Designated Inspector who give the Practical test to qualified persons. ITS EASY!

58

Maintenance Department

100 Hr. Inspection Procedure—Engine and Propellers

1. Compression check L & R engine. Record each cylinder pressure.
2. Inspect and correct exhaust system discrepancy if any.
3. Drain oil sump and check for metal and re safety.
4. Check main oil screen for metal and re safety.
5. C-2 fuel screen (replace screen it rusted or corroded.)

   Procedure for installing fuel screen:
   a. Use new gasket
   b. install screen
   c. replace screen cover and leave thumb screw loose.
      safety wire lo-se shaft to other side of screen cover.
      then secure thumb screw and safety it. (finger tight plus one turn)

6. Inspect general condition of engine and correct any discrepancy.
7. Grease Hartzel Prop. at every 100 hr. inspection.

   Three bladed prop. procedure
   a. remove spinner
   b. check for oil leaks and loose blades
   c. one each prop. blade remove(one)grease fitting to allow old
      grease to escape
   d. grease prop. and replace fitting
   e. replace spinner, observing index marks or aligning marks

f.   spinner must be replaced in the same position it was in before
       it was remove "MANDATORY"
(A&E ONLY)g. dress prop blade as necessary with and feeling
       "Take it easy"—don't file the blade off.

HAMILTON STANDARD PROPELLERS

1. check for oil leaks (A&E ONLY)
2. dress prop. blades as necessary with caution and feeling "take it easy"
    Don't file off a blades.
3. clean blade if not (treated) with solvent and 400 wet/dry paper" ask
    it you don't know the difference"
4. clean off rust and paint hub with aluminum paint, (spray can) watch
    you don't get spray on engine and cowling

Follow inspection form for engines and complete 100 hr. inspection.
This sheet will be revised as necessary to ensure the best possible
procedures which will guarantee the safety of our passengers.

NOTE:

Tools required for post flight inspection
1. 9/16 open end wrench
2. medium size screwdriver
3. flash light
4. rag

(NO EXCEPTIONS)

Maximum allowed time for engine 100 hr. inspection, 3 hours. See
Supervisor if you have any problems.

59

Daily Flight Inspection

1. Chalk aircraft
2. Check mag. sw. (off) (Battery sw. off, Check condition of battery by turning one sw. on and off at a time) Cockpit and cabin for broken or frayed seat belts, seats, seat covers for condition.
3. Check condition of tire and wheels. Check for condition of through bolts on wheels. Trio-gear, tire pressure 65 psi. (MAINS) 45 psi. nose WHEEL (57) all tire pressure 45 psi.
4. Check wheel wells for hydraulic leaks, engine oil leaks, fuel and fuel stains general condition of structure bung springs.
5. Check struts for hydraulic leaks—strut height 4 inch.
6. Nose shimmy damperner for leaks and check for security of damperner. (IF IT LEAKES, REPLACE IT.)
7. Check condition of brakes, leaking hydraulic fluid, worn pucks, etc.
8. Check engine per previos instructions. Check exhaust pipes for loose nuts and missing studs. Check for oil leaks, fuel leaks (stains). Check for oil leaks on cylinder at barrel and head attach point (investigate all fuel and oil leaks.
9. Check prop. blades for nicks, leading edge for roughness (A&E to dress prop blades only.)
10. Check all 1500 psi lines, must be replaced if frayed or if steel braid is showing. (Brake lines, etc.)
11. Check condition of engine cowling and for (mirror) on left cowl inboard (all aircraft except 57)
12. Check condition of flight controls, for holes and freedom of movement (fabric).

13. Check condition of cabin door chain and locking mechanism.
14. Check fuselage for structure, skin, condition of paint, any damage marks.
15. Check all windows—cockpit, cabin for crazing, cracks, dirt and etc.
16. Check aircraft for cleanness interior and exterior.

NOTE: REPORT ANY DISCREPANCY TO THE SUPERVISOR, ENTER IT IN THE DAILY WORK SHEET AND CORRECT ASAP THE SAME NIGHT.

NOTE: Tools required to perform post flight inspection.

A. 9/16 open end wrench
B. medium size screwdriver
C. flash light
d. rag

60

Starting Aircraft Engine

1. FIRE GUARD STAND BY
2. CHECK AIRCRAFT PRIOR TO START
3. PARKING BRAKE SET
4. FUEL SELECTOR ON MAIN TANK
5. BATTRIES ON—AFTER DARK NAVIGATION LIGHT ON, ROTATING BEACON ON
6. Mixture levers Full Rich (FOWARD).
7. Engine Selectors SW. to desired engine L or R.
8. Pump throttle lever 5 times and return to idle position.
9. Fire guard clears PROP.
10. Press starter button (SW) and open throttle (full forward) until 6 blades have pasted then return throttle lever to idle position while continually pressing starter button (SW).
11. Magneto SW. on.
12. After engine starts increase RPM to 1000-warm up engine and perform desired test-etc.
13. All switches—fuel selectors off—(parking brake set). During rainy months more fuel may be required to start engine simply pump throttle lever a few more times 3-7 and repeat10-12, If aircraft sets 24 hr. or more turn propeller by hand at least 12 blades (mag. SW off) before starting engine to check for hydraulic cylinders.
14. FIRE GUARD MANDATORY

61

Read It And Believe It

Daily post flight inspection procedures and necessary equipment which is required to be in the mechanic possession while on the ramp performing post flight inspections and NOT locked up in his tool box.

This list of tools, procedures and methods ARE MANDATORY THERE ARE NO EXCEPTIONS

1. 9/16 open end wrench. Check security of every cylinder hold down nut that you can get a wrench on. IF IT MOVES IT IS LOOSE!!! Retorque all nuts (315" lb) on that cylinder. If loose nuts are found on more than one CYLINDER, all CYLINDERS will be retorted. This means all 9 cylinders GET AN EARLY START SHE FLIES TOMORROW.
2. Screw driver (medium size)
   A. Check security of cowling
   B. Check for loose DUZE FASTENERS
3. Flashlight—to see with
4. RAG—Wipe oil off cowling (INVESTIGATE ALL OIL LEAKS)

62

Emergency Code

A. DAY TIME-PERSON WAVING A FLAG
B. NIGHT TIME-PERSON GOES TO THE AIRCRAFT NEAREST THE ACCIDENT AND TURNS THE ROTATING BEACON ON THAT AIRCRAFT.

INITAL, ALL MECH.

1.
2.
3.

HUTCH AIR

I was hired as the Director of Maintenence it work well for a while. Then they hire a retired airline pilot as the Director of Operation and we got along for awhile, one day he complaining about the fact, he had no responsibly on his job so he wanted mine.

The next day he came to work and told me, the 208s has first priority and the 170 any time after. A couple days later I had the crew working on the 170s and he came out of the office screaming why the crew was working on the 170s I said, because the 208s are all flying.

If you like the mechanic can hang by one hand and tighten a bolts with the other while the 208s are at 5000' but I THOUGHT PRIORTY meant while they were on the ground.

The crew started to laught, he flew in to the office and yelled (I quit) and left. About two week later the company closed the doors.

In 1989 I went back to school to finish my Ph.D. I had started working on it in 1974 and completed it in DEC. 1992.

"THE MORE YOU KNOW, THE FURTHER YOU GO"

63

About The Author

NAME   JACK L. DOWD PH.D.
Place & Date of Birth   OTTUMWA, IA. 6/1/32
Service (NAVY)   JUNE 1951-JUNE 1955
Education   OTTUMWA HIGH SCHOOL
                                     NORTHROP AERO. INSTITUTE
                                     UNIVERSITY OF HAWAII
                                     HONOLULU UNIVERSITY

64

Bibliograpy

I have no list or source of information, such as books pamphlets from which to draw from. I drew from my own experience while employed as an AIRCRAFT MECHANIC and with an INSPECTION AUTHORIZATION which I held for 6 years.

'THIS WORK IS FICTION BASED ON FACT

1. Airframe & Powerplant Mechanics AC-12A Page 18, Cylinders.
2. Acceptable Methods, Technique and Practices EA-AC 43.13-1A & A2 Page 271 2ND Col. 2ND Paragraph
3. Airframe & Powerplant Mechanic Airframe Handbook Page 25
4. Honolulu Star Bulletin Saturday, November 21, Page 1, A-3

SYNOPSIS

The title PLANE TALK may seem to be naïve. I say this because I didn't have any experience working on Airplanes. I receive my License 9/15/58 and I continued with the Jet Engine and Overhaul Course right after.

I didn't have much experience up until I started working with Panorama Air Tour.

We had three mechanics with License and three with out. We had 18 Beech 18's so we had a pair of mechanics for each set of six.

The setup we had worked out very well. The Fellas work well together and we could send the guys to get there license after the worked 1and ½ years on each Engine and Airframe completed. It was a long time, but some of then got they license and went to work for the Major AIRLINES.

I thing I did was to secure my self as the Lead mechanic I felt since I would have to go rescue our Airplanes from the outer Island. Since my life was put on the line every time we went to the outer Island I wanted to be the one who call the shots.

We had some Magnetos and some Engine Changes. Then one day I devise a working Preventive Maintenance Routine and all our problems were solve.

AUTHORS BIOGRAPHY

Jack L. Dowd Ph.D.
1026-1026 Haihanupa Street
Ewa Beach, Hawaii 96706
808 6894344 Phone & FAX (call me before you FAX)

Date of BIRTH	6/1/32
Marital Status	Married NUTCRACKER1@HAWAIIANTEL. NET

Education High School:	Ottwmwa High School, Ottumwa Ia.
College:	Leeward Community College, P.C. Oahu Hawaii
University of Hawaii:	Bachelor of Education Degree
	Honolulu University Masters of Arts Degree Education
	Honolulu University Doctor of Philosophy Education

Technical School:	Northrop Aeronautical Institute
Courses:	Airframe & Powerplant Mechanics License No. 1432080
	Jet Engine & Overhaul
	I held the Inspection Authorization from 06/03/77 to 03/31/83

Experience:	1959-1970 General Aviation Maintenance, 12/71-11/79 PANORAMA AIR TOUR.

1981-7/81 Barbers Point Flying Club, 11/81-12/83 Taught Basic Electronics For the Dept. of Education, December 6th, 1983 suffered a stroke, Return to work 10/85-4/86

Scenic Air Tour Director of Maintenance, Honolulu Community College Aircraft Maintenance Instructor 8/87-5/88, Hutch Air Tour Director of Maintenance

3/89-2/90, ALOHA STATE tours & TRANSPORTATION DIRECTOR OF TRAINING/PERSONNEL 3/90-9/92, 10 years in PRINTING, 10 YEARS IN ELECTRONIC.

Why I think my book is especially marketable.

I went through Aviation school and I would like to have had a small book to glance at to show me the way. My book don't tell you how the class is run, but can show you the way that you might want to go.

Nutcracker1@ hawaiiantel.net

www.ingramcontent.com/pod-product-compliance
Lightning Source LLC
Chambersburg PA
CBHW021958170526
45157CB00003B/1045